室内细部设计书系
SERIES OF INTERIOR DETAILED DESIGN

室内设计节点手册
THE ILLUSTRATED HANDBOOK OF INTERIOR DETAILED DESIGN

常用节点
GENERAL DETAILS

第 2 版
2ND EDITION

主编
赵 鲲 朱小斌 周遐德

绘图
陈 琦 刘 敏

图书在版编目（CIP）数据

室内设计节点手册. 常用节点 / 赵鲲，朱小斌，周遐德主编. —2版. —上海：同济大学出版社，2019.5（2021.12 重印）
（室内细部设计书系）
ISBN 978-7-5608-8537-7

Ⅰ. ①室… Ⅱ. ①赵… ②朱… ③周… Ⅲ. ①室内装饰设计-手册 Ⅳ. ① TU238-62

中国版本图书馆 CIP 数据核字（2019）第 081457 号

室内细部设计书系

室内设计节点手册：常用节点（第 2 版）

赵　鲲　朱小斌　周遐德　主编

出 品 人： 华春荣
责任编辑： 吕　炜
助理编辑： 吴世强
责任校对： 徐春莲
装帧设计： 完　颖
装帧制作： 嵇海丰

出版发行：	同济大学出版社 www.tongjipress.com.cn
	（地址：上海市四平路 1239 号　邮编：200092　电话：021-65985622）
经　　销：	全国各地新华书店、建筑书店、网络书店
印　　刷：	上海安枫印务有限公司
开　　本：	889mm×1194mm　1/16
印　　张：	14.25
字　　数：	456 000
版　　次：	2019 年 5 月第 2 版　2021 年 12 月第 4 次印刷
书　　号：	ISBN 978-7-5608-8537-7
定　　价：	139.00 元

版权所有　侵权必究　印装问题　负责调换

"dop 设计"公众号
室内设计领域技术性专业知识分享媒体
目前已经有近 20 万设计师关注

扫码关注公众号,回复关键词
找到你想了解的专业知识

前　言

距离 2017 年 1 月《室内设计节点手册：常用节点》的第一次正式出版已经有两年多的时间了。在这段时间里，我们看到了本书给设计师朋友们带来的帮助，很多设计师因为这本书的内容得到了专业上的提升；也看到了本书给行业带来的影响，涌现出了大量采用新的表达方式来分析技术问题的案例。这些都反映出设计行业对施工图的重视以及设计师对学习的渴望。

同时，我们也收到了很多读者具体的反馈意见和建议，希望本书能进一步升级。所以，在大家呼声比较高的三个方面，我们对原书内容进行了修订：

1. 在节点图纸上增加了尺寸标注

原来我们认为尺寸标注过于固化会误导读者对节点的学习和认知，但是没有尺寸确实会让很多读者无据可依。因此，本次修订增加了大部分节点的控制性尺寸标注，给大家提供一个合理的尺寸参考。

2. 在原有节点内容的基础上新增了 35 个常用节点

由于本书主推的是常用节点，所以在节点的选择上一直比较严格，不会为了堆积数量而去选择并不通用的节点。这次修订所增加的节点严选的都是设计师在工作中会高频使用到的节点。

3. 将本书所涉及的节点图纸以电子版的形式公开

这应该是所有读者最大的愿望了。为了方便大家更进一步地学习和使用室内常用节点，针对此次购买《室内设计节点手册：常用节点（第 2 版）》的朋友，我们会给出下载路径，您可以从"设计得到"网站（www.shejidedao.com）下载到书中所有节点图纸的 CAD 电子文件。

4. 提供了所有节点图的 GIF 格式动图

通过扫描书中的二维码，读者可以看到三维模型的分解步骤动图。

此次对《室内设计节点手册：常用节点》进行修订，我们依然拿出最开放的心态，顺应时代的发展，为设计师朋友在工作和学习上提供最切实的帮助。在此，也非常感谢广大读者给予节点手册和 dop 的一贯支持！

赵　鲲
2019 年 4 月

005　　前言

006　　目录

015　　1　墙面工艺节点

016　　瓷砖/石材湿贴墙面　|轻质砖墙体|
016　　瓷砖/石材湿贴墙面　|钢架墙体|
018　　涂料墙面　|轻质砖墙体|
018　　涂料墙面　|轻钢龙骨墙体|
020　　石材干挂墙面　|剪力墙/柱子|
022　　石材干挂墙面　|轻质砖墙体|
024　　GRG/GRC挂板墙面　|剪力墙/柱子|
026　　GRG/GRC挂板墙面　|轻质砖墙体|
028　　木饰面挂板墙面　|轻质砖墙体|
030　　木饰面挂板墙面　|轻钢龙骨墙体|
032　　木饰面粘贴墙面　|轻质砖墙体|
032　　木饰面粘贴墙面　|轻钢龙骨墙体|
034　　金属薄板粘贴墙面　|轻质砖墙体|
034　　金属薄板粘贴墙面　|轻钢龙骨墙体|
036　　金属挂板墙面　|轻质砖墙体|
036　　金属挂板墙面　|轻钢龙骨墙体|
038　　玻璃饰面墙面　|轻质砖墙体|
038　　玻璃饰面墙面　|轻钢龙骨墙体|
040　　壁纸饰面墙面　|轻质砖墙体|
040　　壁纸饰面墙面　|轻钢龙骨墙体|
042　　软包饰面墙面　|轻质砖墙体|
042　　软包饰面墙面　|轻钢龙骨墙体|

044	硬包饰面墙面	｜轻质砖墙体｜
044	硬包饰面墙面	｜轻钢龙骨墙体｜
046	木质吸声板墙面	｜轻质砖墙体｜
046	木质吸声板墙面	｜轻钢龙骨墙体｜
048	玻璃隔墙	
050	轻钢龙骨墙一	
050	轻钢龙骨墙二	
052	钢架墙一	
052	钢架墙二	
054	轻质墙	
056	幕墙 - 轻钢龙骨隔墙交接一	
056	幕墙 - 轻钢龙骨隔墙交接二	
058	窗台板一	
058	窗台板二	
060	挂墙式马桶	
062	挂墙式小便器	
064	暗水箱蹲便器	
066	侧拉式防火卷帘	

069　2　吊顶工艺节点

070　石膏板吊顶　|贴顶式|
072　石膏板吊顶　|悬吊式|
074　石膏板吊顶　|卡式承载龙骨|
076　石膏板吊顶　|高低差造型|
078　石膏板吊顶　|阴角石膏线条 / 顶面石膏线条|
080　石膏板吊顶　|常规灯槽造型|
082　石膏板吊顶　|带石膏线灯槽造型|
084　石膏板吊顶　|弧形石膏线灯槽造型|
086　石膏板吊顶　|靠墙风口带灯槽造型|
088　石膏板吊顶　|灯槽带风口造型|
090　吊顶　|顶面墙角留缝造型|
092　石膏板吊顶　|顶面留缝造型|
094　明装式窗帘盒天花
096　暗装式窗帘盒天花
098　明装式窗帘盒天花　|低于窗户|
100　暗装式窗帘盒天花　|低于窗户|
102　暗装式投影幕布　|靠墙|
104　暗装式投影幕布　|居中|
106　升降投影仪
108　嵌入式顶花洒
110　涂料顶面与涂料墙面交接天花
112　涂料顶面与石材墙面交接天花

114	矩形金属格栅天花
116	圆形金属格栅天花
118	木饰面吊顶天花
120	软膜吊顶天花
122	亚克力吊顶天花
124	伸缩缝工艺天花
126	成品检修口天花
128	反支撑工艺天花一
130	反支撑工艺天花二
132	挡烟垂壁天花
134	可升降挡烟垂壁天花
136	单轨钢制防火卷帘
138	双轨无机布防火卷帘

141　3　地坪工艺节点

142	石材/瓷砖地坪　｜干铺法｜
142	水磨石地坪
144	木地板地坪　｜混凝土基层｜
144	木地板地坪　｜木龙骨基层｜
146	防腐木地坪
146	环氧地坪
148	架空地板地坪
148	玻璃地坪
150	块毯地坪
150	满铺地毯地坪
152	塑胶地板地坪
152	地暖地坪
154	砌筑地台
154	钢架地台
156	石材 - 木地板交接地坪
156	石材 - 满铺地毯交接地坪
158	石材 - 除泥垫交接地坪
158	木地板 - 满铺地毯交接地坪
160	地坪 - 幕墙收口　｜高｜
162	地坪 - 幕墙收口　｜平｜
164	卫生间淋浴房挡水槛地坪（铺法一）
164	卫生间淋浴房挡水槛地坪（铺法二）

目录 CONTENTS

166　卫生间地漏　|明装式|
166　卫生间地漏　|隐藏式|
168　卫生间门槛石地坪（铺法一）
168　卫生间门槛石地坪（铺法二）
170　卫生间玻璃隔断墙面收口
170　墙地面防水
172　地面变形缝
172　墙面变形缝
174　泳池隐藏式排水沟
176　泳池明装式排水沟
178　石材踢脚　|凸|
178　石材踢脚　|平|
180　金属踢脚　|凹|
180　金属踢脚　|凸|
182　石材踏步　|混凝土楼梯|
182　石材踏步　|钢结构楼梯|
184　木地板踏步　|混凝土楼梯|
184　木地板踏步　|钢结构楼梯|
186　地毯踏步　|混凝土楼梯|
186　地毯踏步　|钢结构楼梯|
188　石材踏步　|有灯带|
188　木地板踏步　|有灯带|

191　**4　门工艺节点**

192　地弹簧玻璃门
194　玻璃铰链门　|固定玻璃|
196　玻璃铰链门　|固定墙面|
198　双开门
200　单开门
202　暗藏移门
204　贴墙明装移门
206　同向联动移门
208　电动玻璃移门

210	不锈钢电梯门套
210	石材电梯门套
212	钢制单开防火门
214	钢制子母防火门
216	木质单开防火门
218	木质子母防火门
220	常开防火门
222	装饰暗门　｜双道门｜
224	石材暗门

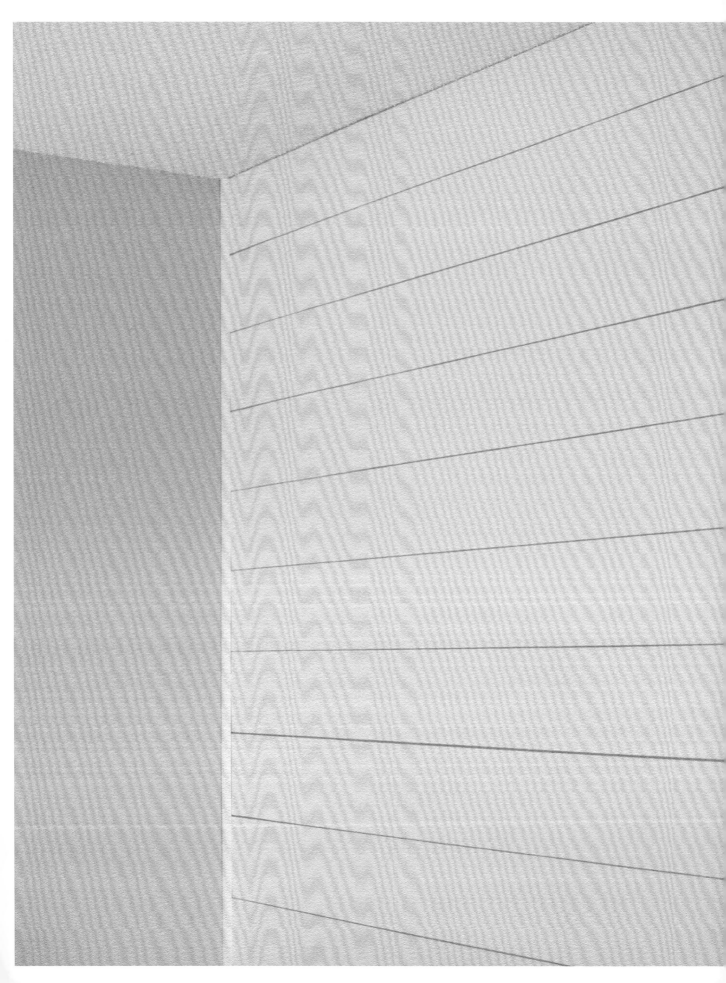

1

墙面工艺节点
DETAILS OF WALL PROCESSING TECHNIQUES

　　本篇的主要内容是几种常见的墙面材料做法和隔墙做法，涉及的材料有瓷砖、石材、GRG（玻璃纤维增强石膏成型材料）、木饰面、金属、玻璃，工艺有干挂、湿贴等，涉及的隔墙有轻质砖墙、轻钢龙骨隔墙、钢架墙、玻璃隔墙等。

　　墙面采用不同的材料就需要不同的工艺做法，这会造成不同的墙体完成面厚度。完成面厚度会对空间的尺度和造型产生较大的影响。因此，在设计过程中，首先要清楚所用装饰材料的属性和规格，其次要了解工艺，比如石材的厚度是多少，安装完厚度是多少，有了这些基本判断就可以计算出完成面尺寸，从而让我们的设计更加准确、可行。

瓷砖/石材湿贴墙面　|轻质砖墙体|
Ceramic Tile/Stone Adhesive-cladding Wall　| Light-weight Brick Wall |

016P ↗　017P ↗

重点 / KEY POINTS

粉刷层厚度一般 2 cm 左右为宜，专用黏结剂厚度一般在 3~5 mm。具体根据饰面材料厚度及属性而定。

The width of paint layer is about 2 cm. And the special adhesive is 3~5 mm. It should be adjusted according to the width and feature of the finish material.

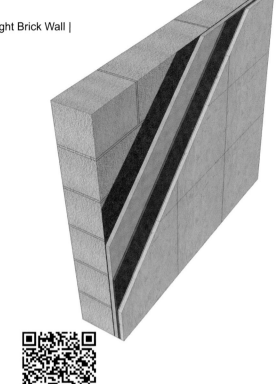

微信扫码，了解更多关于"石材铺贴"的知识

微信扫码，观看"瓷砖/石材湿贴墙面|轻质砖墙体|"三维节点动图

微信扫码，观看"瓷砖/石材湿贴墙面|钢架墙体|"三维节点动图

重点 / KEY POINTS

钢架墙特征：墙身薄，墙体轻，施工方便快捷。

Features of steel truss wall: thin, Light-weight, easy and quick installation.

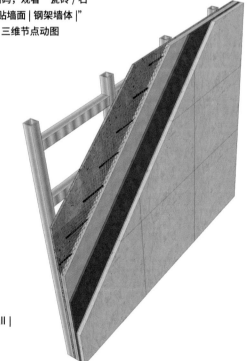

瓷砖/石材湿贴墙面　|钢架墙体|
Ceramic Tile/Stone Adhesive-cladding Wall　| Steel Truss Wall |

016P ↘　017P ↘

三维图　PERSPECTIVE

墙面工艺节点　DETAILS OF WALL PROCESSING TECHNIQUES

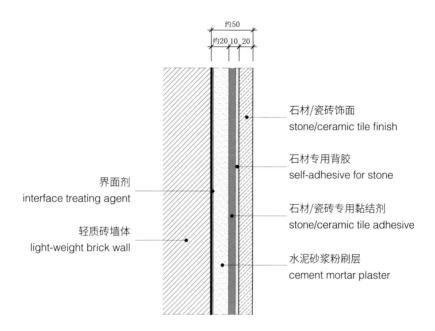

比例　scale: 1:5

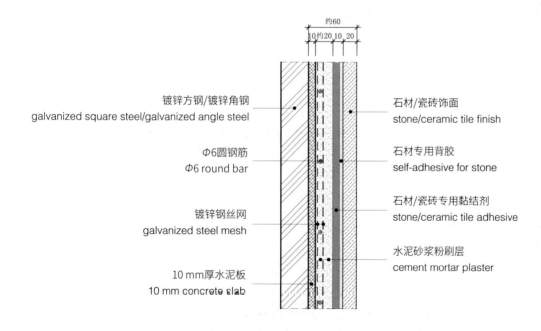

涂料墙面　| 轻质砖墙体 |
Paint Wall　| Light-weight Brick Wall |

018P ↗　019P ↗

重点 / KEY POINTS

　　轻质砖墙砌筑好后用水泥砂浆粉刷，腻子找平，刷乳胶漆即可。

After the Light-weight brick wall being mortared and constructed, paint it with cement mortar, then level it with putty, and finally apply a coat of emulsion paint.

微信扫码，了解更多关于"涂料、裱糊工艺"的知识

微信扫码，观看"涂料墙面 | 轻质砖墙体 |"三维节点动图

重点 / KEY POINTS

　　底层与面层石膏板必须错缝拼接安装；石膏板与石膏板交接缝隙以 5 mm 左右为宜，然后采用腻子填缝，贴上绷带后进行批嵌。

The base layer and surface layer of gypsum board should be installed by stitching joint. The gap between gypsum board should be 5 mm around. Fill the gap with putty, affix bandage and fill with putty again.

微信扫码，观看"涂料墙面 | 轻钢龙骨墙体 |"三维节点动图

涂料墙面　| 轻钢龙骨墙体 |
Paint Wall　| Light-guage Steel Framing Wall |

018P ↘　019P ↘

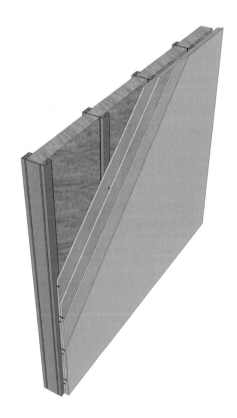
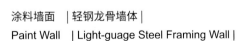

墙面工艺节点　DETAILS OF WALL PROCESSING TECHNIQUES

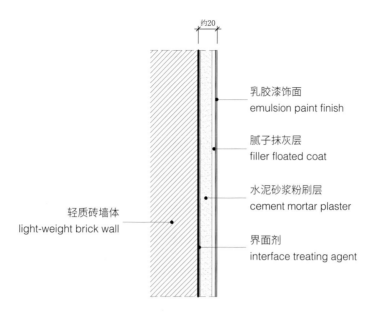

乳胶漆饰面
emulsion paint finish

腻子抹灰层
filler floated coat

水泥砂浆粉刷层
cement mortar plaster

界面剂
interface treating agent

轻质砖墙体
light-weight brick wall

比例　scale: 1:5

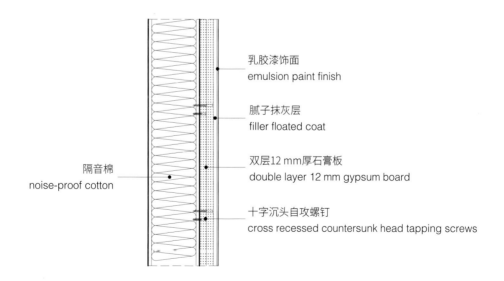

乳胶漆饰面
emulsion paint finish

腻子抹灰层
filler floated coat

双层12 mm厚石膏板
double layer 12 mm gypsum board

十字沉头自攻螺钉
cross recessed countersunk head tapping screws

隔音棉
noise-proof cotton

节点图　DETAIL

石材干挂墙面　|剪力墙 / 柱子 |
Dry Stone Fixing Wall　　|Shear Wall/Pillar |

020P / 021P

重点 / KEY POINTS

剪力墙、柱子一般为钢筋混凝土结构，而只有这类结构才可以直接采用膨胀螺栓固定角钢，然后进行石材干挂。

Generally, the shear wall and pillar are made of reinforced concrete. Expansion bolt can be used to fixing the angle steel only with this structure.

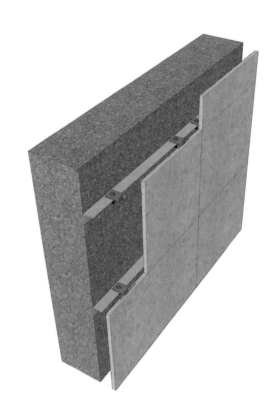

微信扫码，了解更多关于"石材干挂"的知识

微信扫码，观看"石材干挂墙面 | 剪力墙 / 柱子 |"三维节点动图

三维图　PERSPECTIVE

墙面工艺节点　DETAILS OF WALL PROCESSING TECHNIQUES

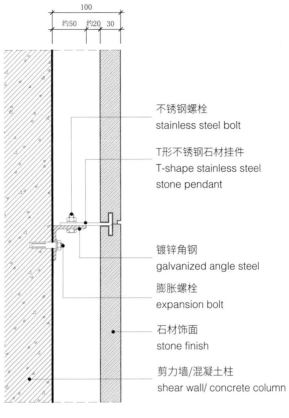

纵剖面　Longitudinal Section

比例　scale: 1:5

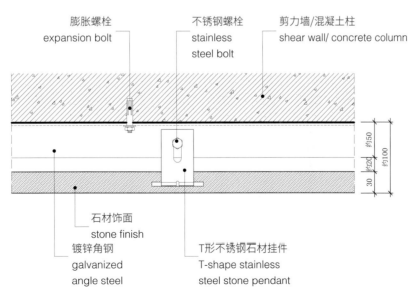

横剖面　Cross Section

节点图　DETAIL

石材干挂墙面　|轻质砖墙体|
Dry Stone Fixing Wall | Light-weight Brick Wall |

022P / 023P

重点 / KEY POINTS

轻质砖墙因无法像剪力墙、柱子一样直接进行石材干挂，所以一般采用独立钢架结构干挂，钢架一般"顶天立地"固定，在建筑圈梁处会进行加固处理。

For the reason that stone cannot be fixed to Light-weight brick wall like shear wall and pillar, independent steel structure is always adopted for stone cladding fixing. The steel structure will be fixed to the ceiling and the floor, and be reinforced in the gird.

微信扫码，观看"石材干挂墙面|轻质砖墙体|"
三维节点动图

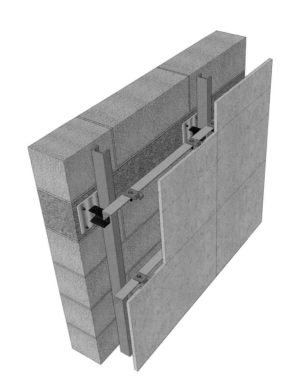

三维图　PERSPECTIVE

墙面工艺节点　DETAILS OF WALL PROCESSING TECHNIQUES

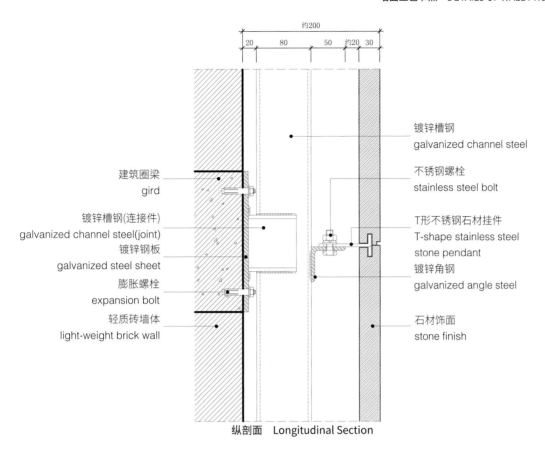

纵剖面　Longitudinal Section

比例　scale: 1:5

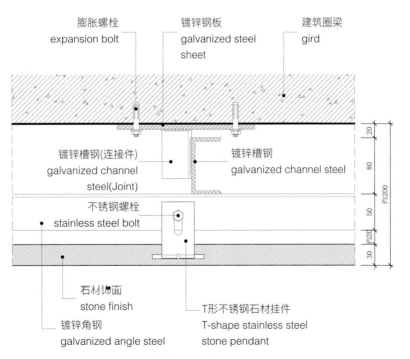

横剖面　Cross Section

节点图　DETAIL

GRG/GRC 挂板墙面　　|剪力墙 / 柱子 |
GRG/GRC Cladding Panel Wall　　| Shear Wall/Pillar |

024P / 025P

重点 / KEY POINTS

　　GRG：加强纤维石膏；GRC：加强纤维水泥。它们的施工方式与石材干挂类似，只不过其挂件在定制加工时已进行预埋，与材料连成一体。

GRC: reinforced fiber gypsum; GRC: reinforced fiber cement. The technique of this material is similar to dry stone fixing. While the anchor is already embedded in the material when prefabricated in the plant.

微信扫码，了解更多关于"GRG 制作、安装"的知识

微信扫码，观看"GRG/GRC 挂板墙面 | 剪力墙 / 柱子 |"三维节点动图

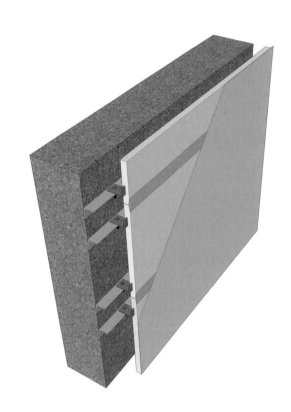

三维图　PERSPECTIVE

墙面工艺节点　DETAILS OF WALL PROCESSING TECHNIQUES

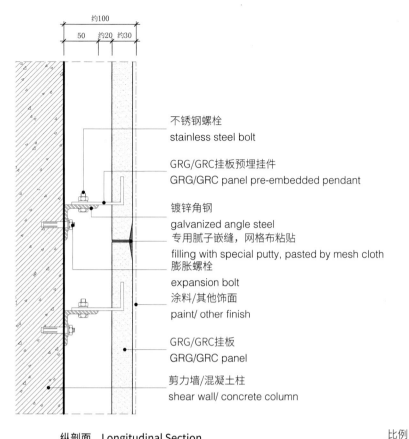

纵剖面　Longitudinal Section

比例　scale: 1:5

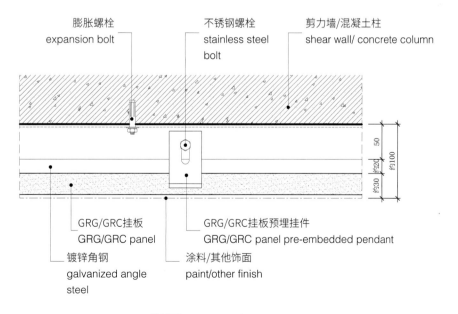

横剖面　Cross Section

节点图　DETAIL

025

GRG/GRC 挂板墙面　|轻质砖墙体 |
GRG/GRC Cladding Panel Wall　| Light-weight Brick Wall |

026P / 027P

重点 / KEY POINTS

　　GRG 更多应用在室内，用来处理大型弧面、曲面造型；GRC 更多应用于室外，如建筑外立面、雕塑等。

GRG is generally applied in interior space. It is often used to make the shape of large arc or curved space; GRC is more often used in exterior space, i.e. architectural elevation, sculpture, etc.

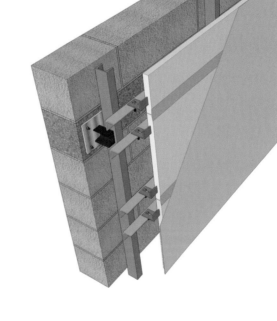

微信扫码，观看"GRG/GRC
挂板墙面 | 轻质砖墙体 |"
三维节点动图

三维图　PERSPECTIVE

墙面工艺节点　DETAILS OF WALL PROCESSING TECHNIQUES

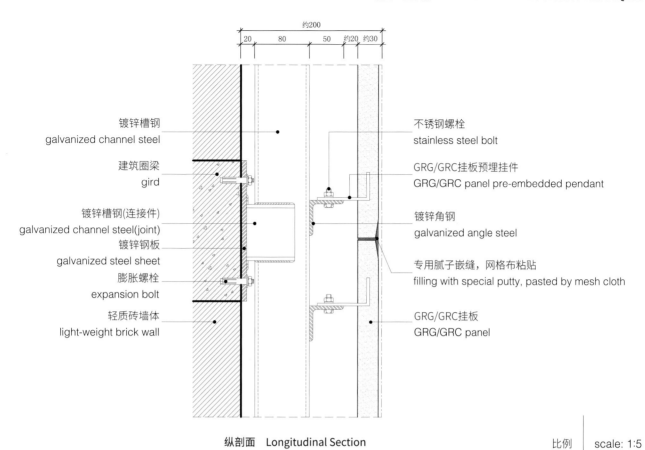

纵剖面　Longitudinal Section

横剖面　Cross Section

比例　scale: 1:5

节点图　DETAIL

木饰面挂板墙面　　|轻质砖墙体|
Wood Finish Panel Wall　|Light-weight Brick Wall|

028P / 029P

重点 / KEY POINTS

在墙面完成面空间较小的情况下可以采用可调节 U 形夹 + 副龙骨的基层处理方式。

Under the scenario that the space is limited after the wall finished, U shape clip + auxiliary joist could be used to adjust the base layer.

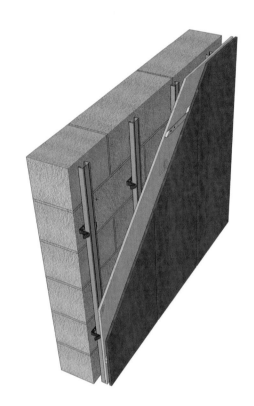

微信扫码，了解更多关于"木饰面"的知识

微信扫码，观看"木饰面挂板墙面|轻质砖墙体|"三维节点动图

墙面工艺节点　DETAILS OF WALL PROCESSING TECHNIQUES

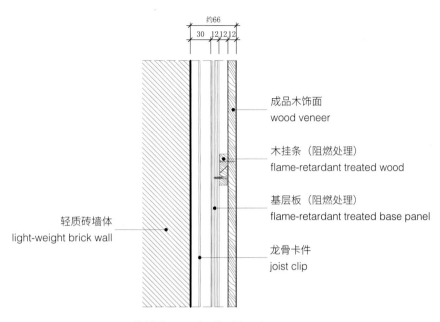

纵剖面　Longitudinal Section

比例　scale: 1:5

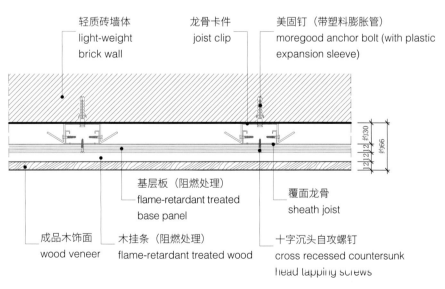

横剖面　Cross Section

节点图　DETAIL

029

木饰面挂板墙面　　|轻钢龙骨墙体|
Wood Veneer Wall　　| Light-guage Steel Framing Wall |

030P / 031P

重点 / KEY POINTS

　　成品木饰面板一般是由 0.6 mm 的木皮 +12 mm 多层板构成的。根据设计要求不同，多层板也可以采用不同的厚度。

Finished wood veneer panel is generally made from 0.6 mm-thick wood veneer and 12 mm-thick plywood. According to design requirement, the thickness of plywood may vary.

微信扫码，观看"木饰面挂板墙面 | 轻钢龙骨墙体 |"三维节点动图

三维图　PERSPECTIVE

墙面工艺节点　DETAILS OF WALL PROCESSING TECHNIQUES

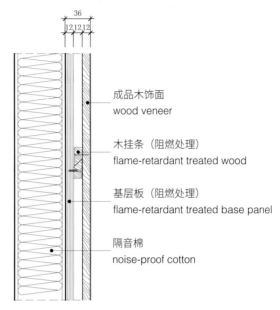

成品木饰面
wood veneer

木挂条（阻燃处理）
flame-retardant treated wood

基层板（阻燃处理）
flame-retardant treated base panel

隔音棉
noise-proof cotton

纵剖面　Longitudinal Section

比例　｜　scale: 1:5

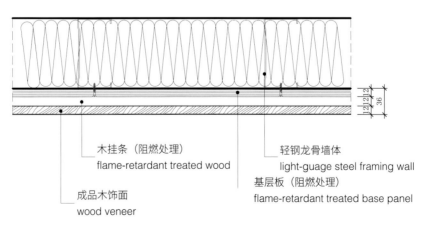

木挂条（阻燃处理）
flame-retardant treated wood

轻钢龙骨墙体
light-guage steel framing wall

成品木饰面
wood veneer

基层板（阻燃处理）
flame-retardant treated base panel

横剖面　Cross Section

木饰面粘贴墙面　|轻质砖墙体|
Wood Veneer Wall　| Light-weight Brick Wall |

032P ↗　033P ↗

重点 / KEY POINTS

小范围的木饰面安装可以采用粘贴式。

If the area of wood veneer is limited, adhesive can be used for its installation.

微信扫码，观看"木饰面粘贴墙面|轻质砖墙体|"三维节点动图

微信扫码，观看"木木饰面粘贴墙面|轻钢龙骨墙体|"三维节点动图

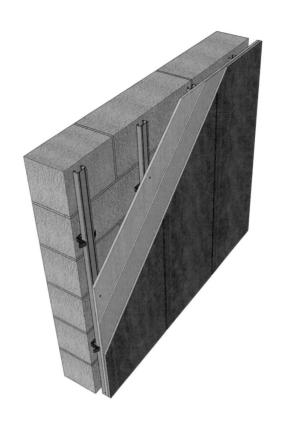

重点 / KEY POINTS

轻钢龙骨墙一定要加吸音棉，以保证隔音效果。

Acoustic absorption layer should be installed to ensure the effect of sound insulation in light-gauge steel framing wall system.

木饰面粘贴墙面　|轻钢龙骨墙体|
Wood Veneer Wall　| Light-gauge Steel Framing Wall |

032P ↘　033P ↘

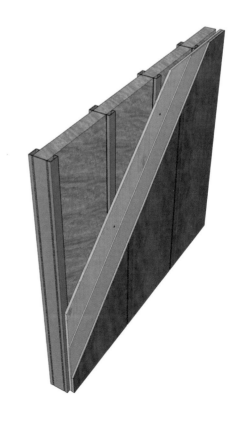

三维图　PERSPECTIVE

墙面工艺节点　DETAILS OF WALL PROCESSING TECHNIQUES

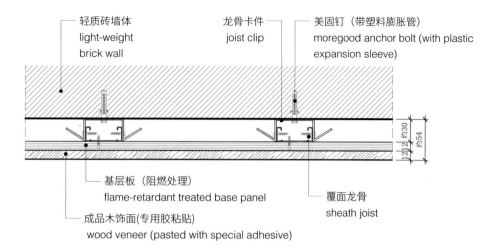

scale: 1:5

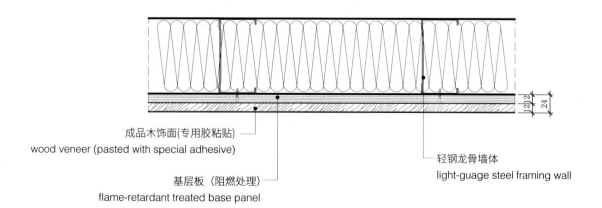

节点图　DETAIL

| 金属薄板粘贴墙面　　| 轻质砖墙体 |
Metal Veneer Wall　　| Light-weight Brick Wall |

034P ↗　035P ↗

重点 / KEY POINTS

　　这里所说的金属饰面板特指不锈钢板，装饰中常用的厚度规格有1 mm，1.2 mm，1.5 mm，2 mm，3 mm 等。

The metal finish board mentioned here is stainless steel board, its common thickness are 1 mm, 1.2 mm, 1.5 mm, 2 mm, 3 mm, etc.

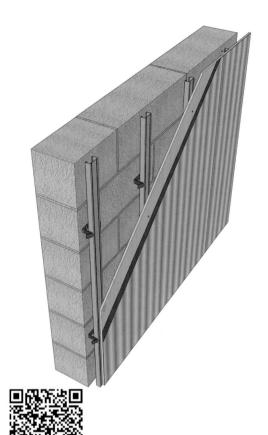

微信扫码，了解更多关于"金属板"的知识

微信扫码，观看"金属薄板粘贴墙面 | 轻质砖墙体 |"三维节点动图

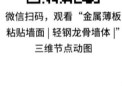

微信扫码，观看"金属薄板粘贴墙面 | 轻钢龙骨墙体 |"三维节点动图

重点 / KEY POINTS

　　粘贴式通常应用于小面积的饰面。

Adhesive is usually applied in a small area of the finish.

| 金属薄板粘贴墙面　　| 轻钢龙骨墙体 |
Metal Veneer Wall　　| Light-gauge Steel Framing Wall |

034P ↘　035P ↘

三维图　PERSPECTIVE

墙面工艺节点　DETAILS OF WALL PROCESSING TECHNIQUES

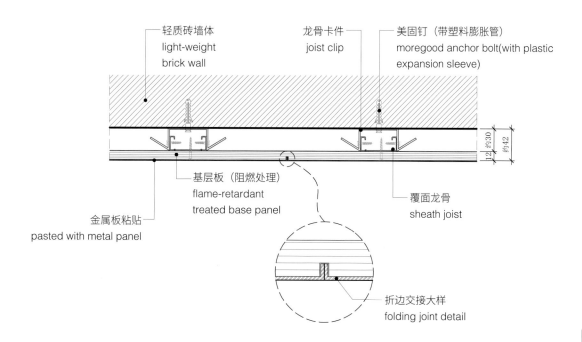

scale: 1:5

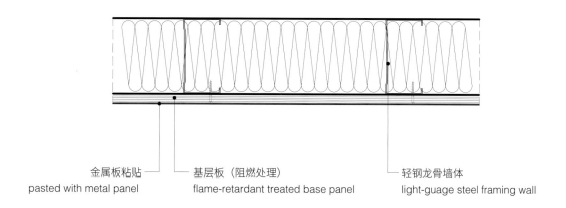

节点图　DETAIL

金属挂板墙面　　|轻质砖墙体 |
Metal Panel Wall　　| Light-weight Brick Wall |

036P ↗　　037P ↗

重点 / KEY POINTS

　　挂板安装通常应用于大面积的饰面，由金属板折边后配合连接件与后部钢结构连接。

Panel cladding always used in large area of the finish installation. The ruffled metal panel is connected with the steel structure behind it by anchors.

微信扫码，观看"金属挂板墙面 | 轻质砖墙体 |"三维节点动图

微信扫码，观看"金属挂板墙面 | 轻钢龙骨墙体 |"三维节点动图

重点 / KEY POINTS

　　金属板面规格较大时，为保证平整度可以采用金属板后加筋的方式，或者可以采用金属板+瓦楞板（蜂窝板）的方式。

When the size of the metal panel is large, reinforced metal panel or metal panel + corrugated board can be used to ensure the smoothness of the finish.

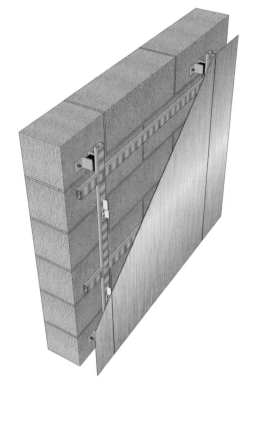

金属挂板墙面　　|轻钢龙骨墙体 |
Metal Panel Wall　　| Light-guage Steel Framing Wall |

036P ↘　　037P ↘

三维图　PERSPECTIVE

墙面工艺节点　DETAILS OF WALL PROCESSING TECHNIQUES

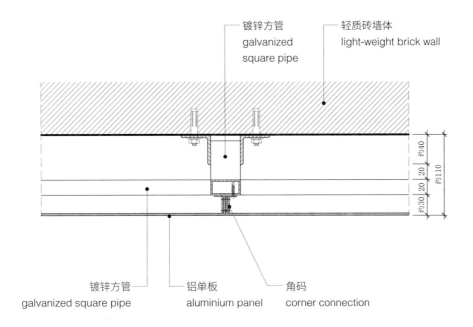

比例　scale: 1:5

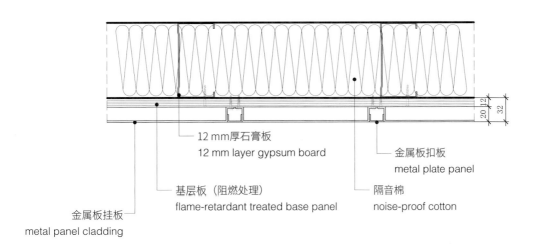

节点图　DETAIL

037

玻璃饰面墙面　|轻质砖墙体|
Glass Finish Wall　| Light-weight Brick Wall |

038P ↗　039P ↗

重点 / KEY POINTS

墙面饰面玻璃厚度一般在6 mm，8 mm，10 mm左右。

The width of glass finish is generally 6 mm, 8 mm, 10 mm, etc.

微信扫码，了解更多关于"玻璃饰面"的知识

微信扫码，观看"玻璃饰面墙面|轻质砖墙体|"三维节点动图

微信扫码，观看"玻璃饰面墙面|轻钢龙骨墙体|"三维节点动图

重点 / KEY POINTS

从安全角度考虑，最好采用钢化玻璃，安装时用双面胶配合玻璃胶固定。

From the sake of safety, it is best to adopt tempered glass and fasten by glass cement with double faced adhesive tape when installed.

玻璃饰面墙面　|轻钢龙骨墙体|
Glass Finish Wall　| Light-gauge Steel Framing Wall |

038P ↘　039P ↘

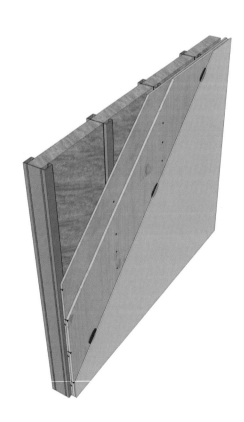

三维图　PERSPECTIVE

墙面工艺节点　DETAILS OF WALL PROCESSING TECHNIQUES

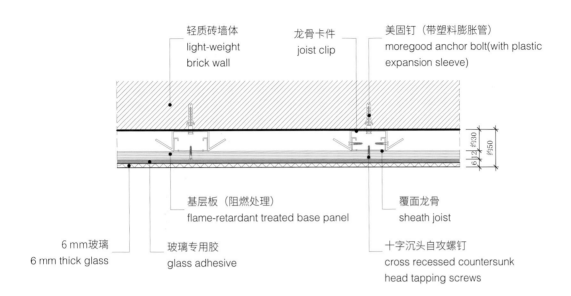

scale: 1:5

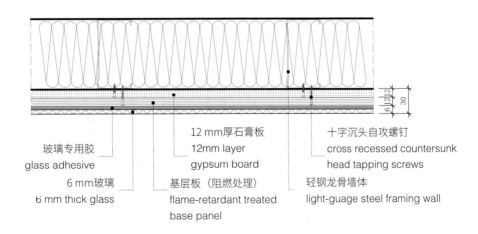

节点图　DETAIL

壁纸饰面墙面　|轻质砖墙体|
Wall Paper Finish Wall　| Light-weight Brick Wall |

040P ↗　041P ↗

重点 / KEY POINTS

墙面处理必须干净、平整、结实、光滑、颜色均匀一致，彻底干燥后施工。

The base should be clean, flat, solid, smooth and color-uniform, process after it dry thoroughly.

微信扫码，观看"壁纸饰面墙面|轻质砖墙体|"三维节点动图

微信扫码，了解更多关于"壁纸、墙布"的知识

微信扫码，观看"壁纸饰面墙面|轻钢龙骨墙体|"三维节点动图

重点 / KEY POINTS

考虑到工艺质量及效果，在有条件的情况下，壁纸铺装的基础还是以石膏板为好。

Considering the quality and effect, if available, the best base of wallpaper is gypsum board.

壁纸饰面墙面　|轻钢龙骨墙体|
Wall Paper Finish Wall　| Light-guage Steel Framing Wall |

040P ↘　041P ↘

墙面工艺节点　DETAILS OF WALL PROCESSING TECHNIQUES

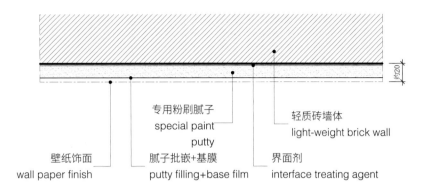

比例 | scale: 1:5

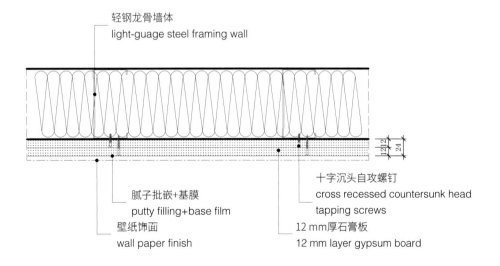

节点图　DETAIL

软包饰面墙面　|轻质砖墙体 |
Upholstered Finish Wall　| Light-weight Brick Wall |

042P ↗　　043P ↗

重点 / KEY POINTS

软包要求基层必须牢固，基层要做抹灰和防潮处理。

The basement layer of upholster must be firm and get plastering and dampproof treatment.

微信扫码，了解更多关于"织物软、硬包"的知识

微信扫码，观看"软包饰面墙面 | 轻质砖墙体 |"三维节点动图

重点 / KEY POINTS

软包的安装方式有很多，胶粘、枪钉固定、魔术贴都可以。

There are many methods of upholster installation, pasting with adhesive, fastening with nailing gun or pasting with magic stick are all available.

微信扫码，观看"软包饰面墙面 | 轻钢龙骨墙体 |"三维节点动图

软包饰面墙面　|轻钢龙骨墙体 |
Upholstered Finish Wall　| Light-guage Steel Framing Wall |

042P ↘　　043P ↘

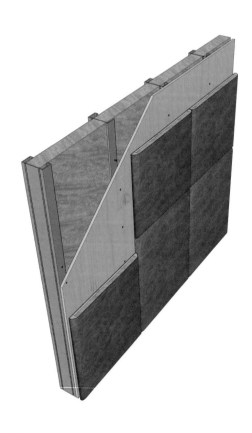

墙面工艺节点 DETAILS OF WALL PROCESSING TECHNIQUES

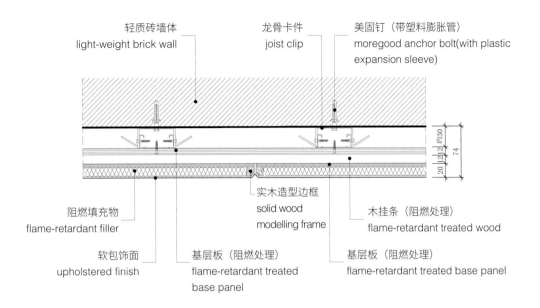

scale: 1:5

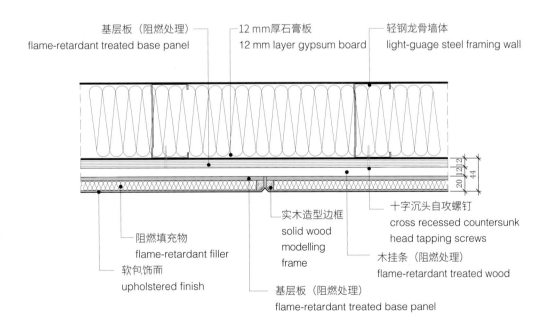

节点图 DETAIL

硬包饰面墙面　　|轻质砖墙体|
Hard Finish Wall　　| Light-weight Brick Wall |

044P ↗　　045P ↗

重点 / KEY POINTS

软包、硬包的区别在于基层板与面层材料之间的填充物，有填充物的为软包，硬包一般无填充物。

The difference between upholstered finish and hard finish is the filler between base board and surface layer of the material. Upholstered finish always has a filler layer, but hard finish doesn't.

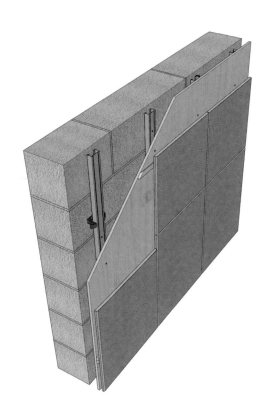

微信扫码，观看"硬包饰面墙面 | 轻质砖墙体 |"三维节点动图

微信扫码，观看"硬包饰面墙面 | 轻钢龙骨墙体 |"三维节点动图

重点 / KEY POINTS

硬包的安装方式和软包相同，可以采用挂板、胶粘、枪钉固定等方式进行安装。

Hard finish is installed the same way as upholstered finish. It can be installed by hanging board, pasting with adhesive and fastening with nailing gun.

硬包饰面墙面　　|轻钢龙骨墙体|
Hard Finish Wall　　| Light-gauge Steel Framing Wall |

044P ↘　　045P ↘

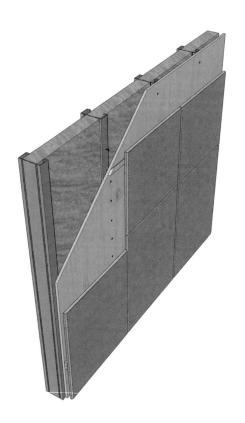

三维图　PERSPECTIVE

墙面工艺节点　DETAILS OF WALL PROCESSING TECHNIQUES

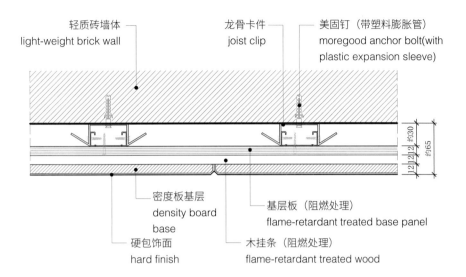

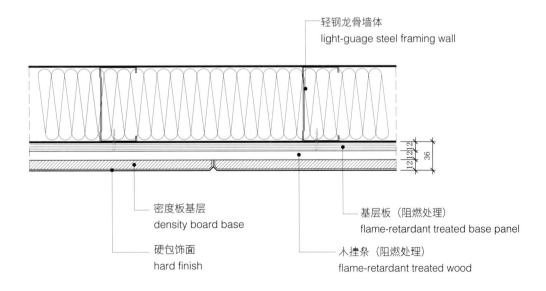

比例　scale: 1:5

节点图　DETAIL

木质吸声板墙面　｜轻质砖墙体｜
Wooden Acoustic Insulation Panel Wall　| Light-weight Brick Wall |

046P ↗　　047P ↗

重点 / KEY POINTS

　　木质吸声板是根据声学原理精致加工而成，由饰面、芯材、吸音薄毡组成。

Wooden acoustic insulation panel is made of the finish, the core and sound-absorbing mat according to acoustic principles.

微信扫码，观看"木质吸声板墙面｜轻质砖墙体｜"三维节点动图

微信扫码，了解更多关于"吸声板"的知识

微信扫码，观看"木质吸声板墙面｜轻钢龙骨墙体｜"三维节点动图

重点 / KEY POINTS

　　吸声板适用于演播室、录音室、剧场、影院等对音质要求较高的空间。

Acoustic insulation panel is used in the spaces which require superb sound quality such as television studio, recording studio, theater and cinema.

木质吸声板墙面　｜轻钢龙骨墙体｜
Wooden Acoustic Insulation Panel Wall | Light-guage Steel Framing Wall |

046P ↘　　047P ↘

三维图　PERSPECTIVE

墙面工艺节点　DETAILS OF WALL PROCESSING TECHNIQUES

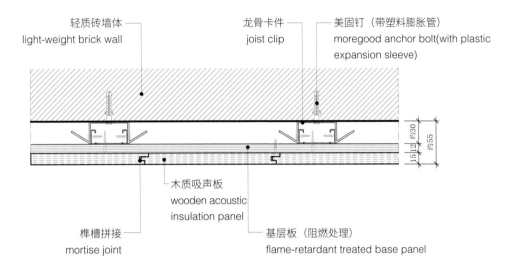

比例 | scale: 1:5

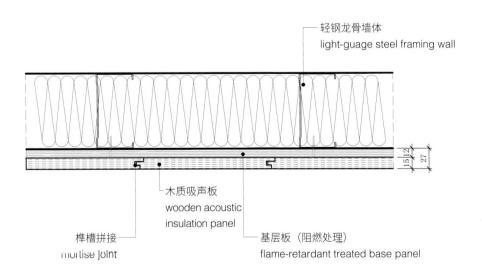

节点图　DETAIL

047

玻璃隔墙
Glass Partition Wall

048P / 049P

重点 / KEY POINTS

玻璃隔墙常用的玻璃厚度有 10 mm 和 12 mm，玻璃必须经过钢化。

Generally, the width of glass that the glass partition wall adopts is 10 mm or 12 mm and the glass must be tempered.

微信扫码，了解更多关于"玻璃隔断"的知识

微信扫码，观看"玻璃隔墙"三维节点动图

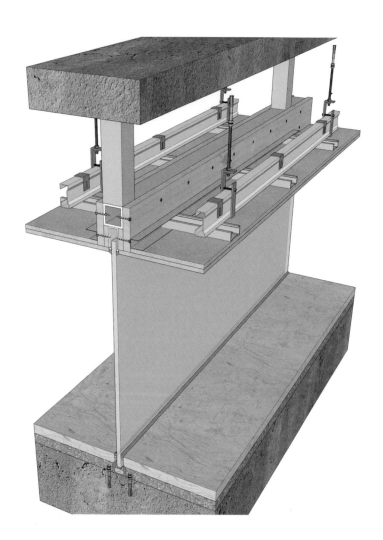

墙面工艺节点　DETAILS OF WALL PROCESSING TECHNIQUES

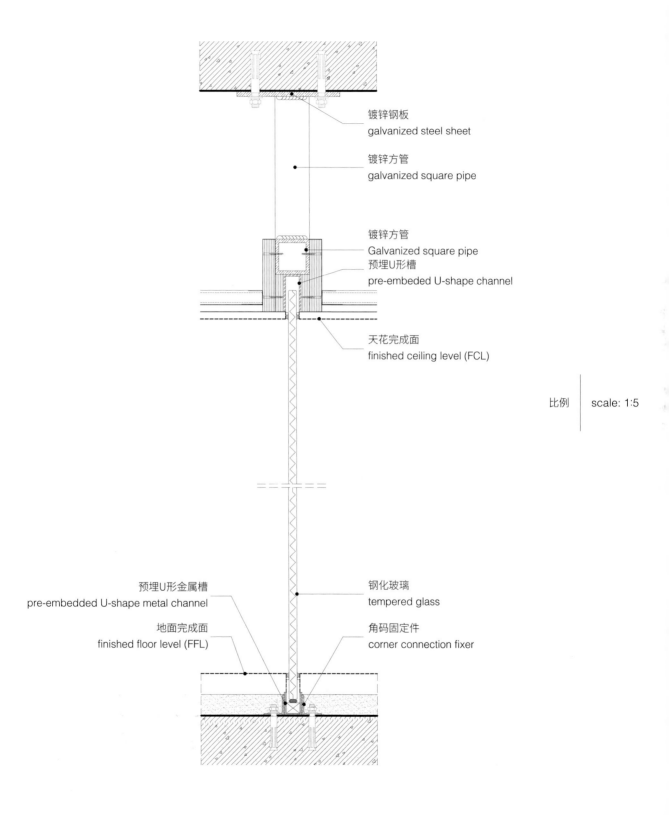

节点图　DETAIL

轻钢龙骨墙一
Light-guage Steel Framing Wall 1

050P ↗   051P ↗

重点 / KEY POINTS

在有防水要求的区域，如：厨房、卫生间等，隔墙下部会设置混凝土导墙，导墙的高度一般在 300 mm 左右。

If the area has to be water-proof, i.e. kitchen, washroom, etc., the lower part of the partition wall will be built as guide wall, and the height of guide wall will be 300 mm.

微信扫码，了解更多关于"轻钢龙骨石膏板隔墙"的知识

微信扫码，观看"轻钢龙骨墙一"三维节点动图

微信扫码，观看"轻钢龙骨墙二"三维节点动图

重点 / KEY POINTS

隔墙龙骨常见有 50 系列、75 系列、100 系列等，以竖向龙骨规格命名。隔墙龙骨主要由沿顶龙骨、沿地龙骨、贯通龙骨、竖向龙骨组成。竖向龙骨的间距一般不宜大于 400 mm。贯通龙骨在隔墙低于 3 m 时安装一道，隔墙为 3~5 m 时，须装两道，也有一些欧美品牌不设置贯通龙骨。

The joist of partition wall has 50 series, 75 series and 100 series, etc. It is named after the size of vertical stud. The joists of partition wall include ceiling joist, floor joist, continuous joist and vertical stud. The vertical span is always less than 400mm. One continuous joist will be installed in the wall lower than 3m and if the partition wall's height is between 3m and 5m, two continuous joists should be installed. Some products of European and American brands do not have continuous joists.

轻钢龙骨墙二
Light-Gauge Steel Framing Wall 2

050P ↘   051P ↘

墙面工艺节点　DETAILS OF WALL PROCESSING TECHNIQUES

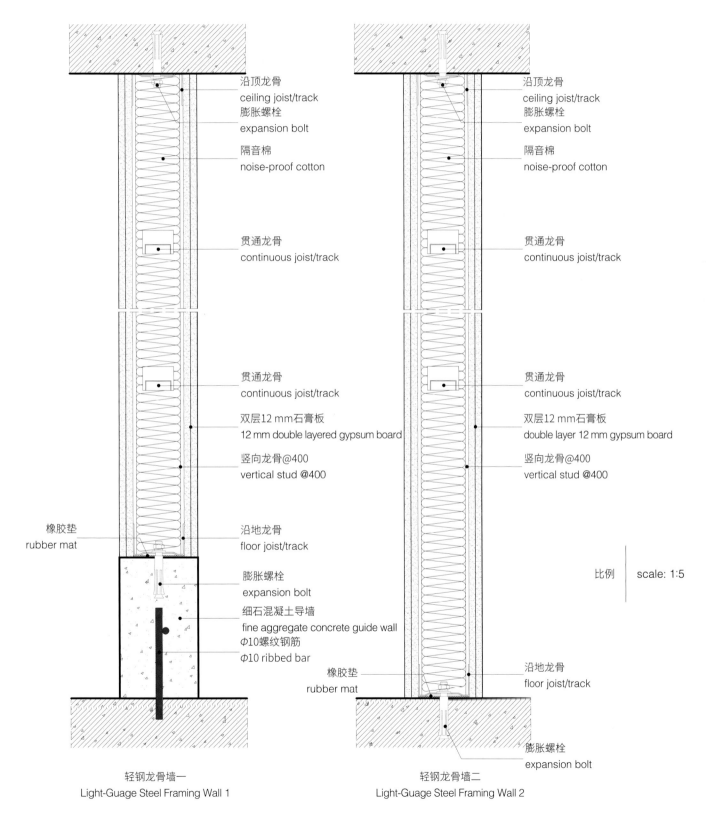

节点图　DETAIL

钢架墙一
Steel Truss Wall 1

052P ↗ 053P ↗

重点 / KEY POINTS

在有防水要求的区域，如：厨房、卫生间等，隔墙下部会设置混凝土导墙，导墙的高度一般在300 mm左右。

If the area has to be water-proof, i.e. kitchen, washroom, etc., the lower part of the partition wall will be built as guide wall, and the height of guide wall will be 300 mm.

微信扫码，了解更多关于"钢架隔墙"的知识

微信扫码，观看"钢架墙一"三维节点动图

微信扫码，观看"钢架墙二"三维节点动图

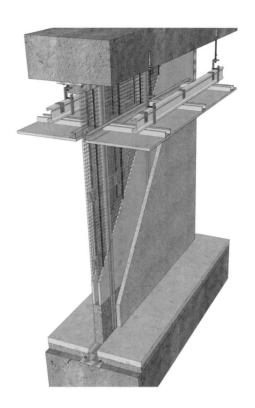

重点 / KEY POINTS

钢架墙的固定需要"顶天立地"：在顶面与地面做预埋件处理，钢架墙再与预埋件焊接。所有构件均需做防锈处理。

The steel truss wall should be fixed to the ceiling and the floor. The wall should be welded to pre-embedded anchors which have been built in the ceiling and floor. All the anchors should be anti-rust treated.

钢架墙二
Steel Truss Wall 2

052P ↘ 053P ↘

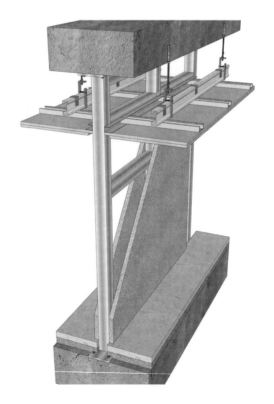

墙面工艺节点　DETAILS OF WALL PROCESSING TECHNIQUES

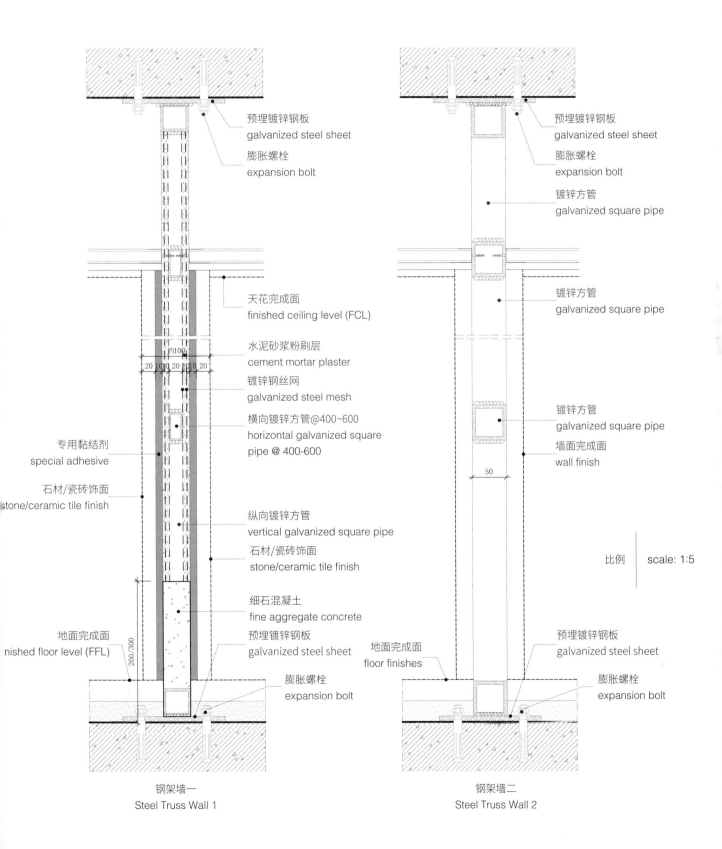

钢架墙一
Steel Truss Wall 1

钢架墙二
Steel Truss Wall 2

节点图　DETAIL

轻质墙
Light-weight Wall

054P / 055P

重点 / KEY POINTS

轻质墙是指用轻质砖或轻质砌块砌筑的墙体，一般由土建单位完成。其特点是质量轻、成本低、施工方便，但是由于强度较低，无法直接在墙体上负载重物。

The Light-weight wall mentioned here is Light-weight brick wall or Light-weight block wall. It is always built by construction company. Its advantage is Light-weight, low-cost and easy to build. But it also has disadvantages—it cannot bear load because of its low strength.

微信扫码，了解更多关于"轻质砖隔墙"的知识

微信扫码，观看"轻质墙"三维节点动图

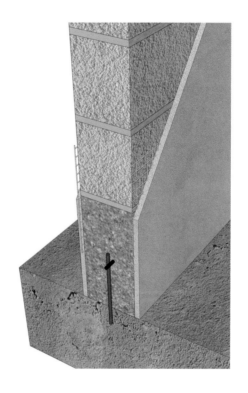

三维图　PERSPECTIVE

墙面工艺节点　DETAILS OF WALL PROCESSING TECHNIQUES

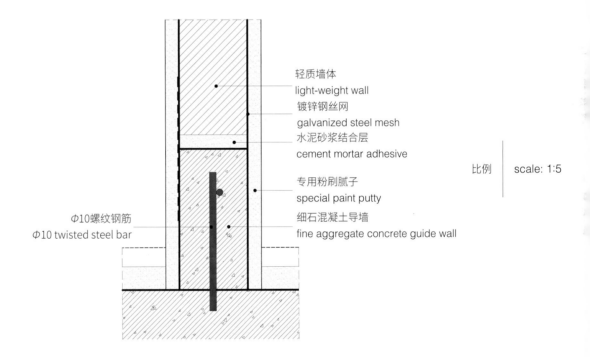

节点图　DETAIL

幕墙 - 轻钢龙骨隔墙交接一
The Connection of Curtain Wall and Light-guage Steel Framing Wall 1

056P ↗   057P ↗

重点 / KEY POINTS

新建隔墙不能和幕墙构造有任何硬性连接，可以采用填充物和打胶密封的方式来处理。

The new partition wall cannot be rigidly connected with the curtain wall structure. It can be treated by filling and sealing.

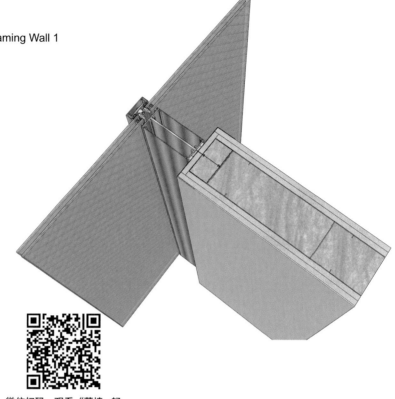

微信扫码，了解更多关于"幕墙和隔墙收口"的知识

微信扫码，观看"幕墙 - 轻钢龙骨隔墙交接一"三维节点动图

微信扫码，观看"幕墙 - 轻钢龙骨隔墙交接二"三维节点动图

重点 / KEY POINTS

对隔音和私密性要求不太高时，可以考虑采用玻璃肋板的方式，设计更多样，施工更便捷。

If the requirements of sound insulation and privacy are not too high, the glass rib board can be considered, in which the design methods are more diversified, and the construction is more convenient.

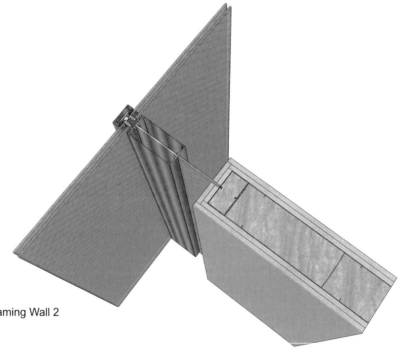

幕墙 - 轻钢龙骨隔墙交接二
The Connection of Curtain Wall and Light-guage Steel Framing Wall 2

056P ↘   057P ↘

三维图　PERSPECTIVE

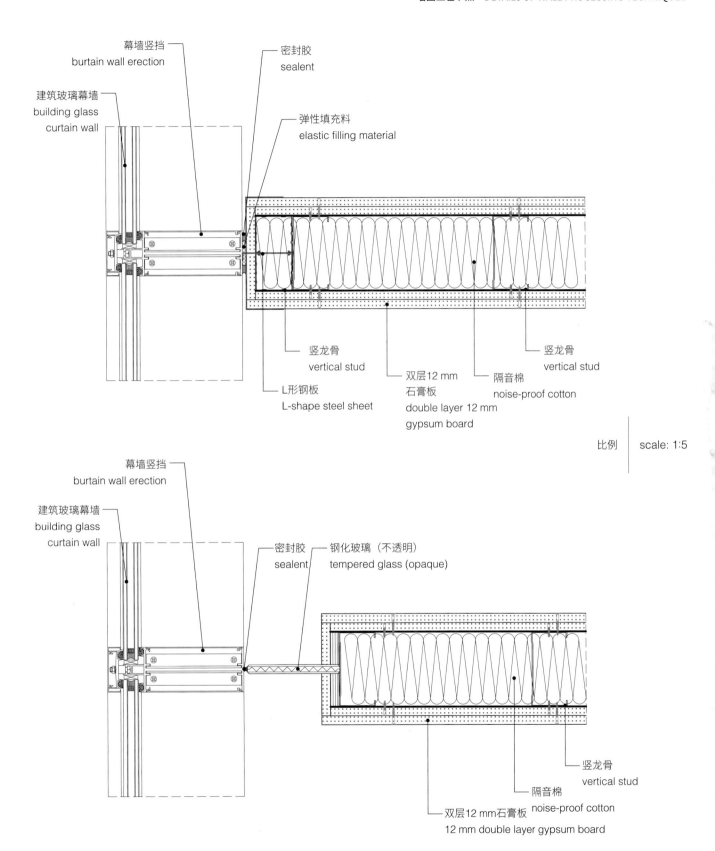

窗台板一
Window Board 1

058P ↗ 059P ↗

重点 / KEY POINTS

窗台土建基础平整度较好，可以直接粘贴；平整度较差或是需要调整窗台板高度的可以考虑使用基层板找平。

If the window foundation has good smoothness, it can be pasted directly. If the window board has poor smoothness or we need to adjust the height of the window board, the leveling of the base plate can be considered.

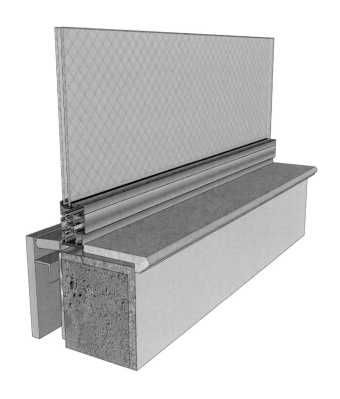

微信扫码，观看"窗台板一"
三维节点动图

微信扫码，观看"窗台板二"
三维节点动图

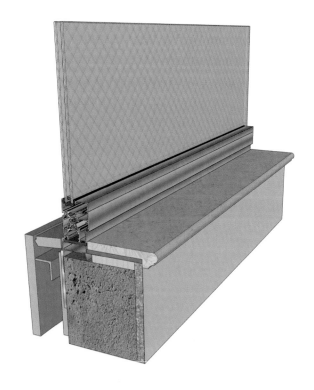

窗台板二
Window Board 2

058P ↘ 059P ↘

墙面工艺节点　DETAILS OF WALL PROCESSING TECHNIQUES

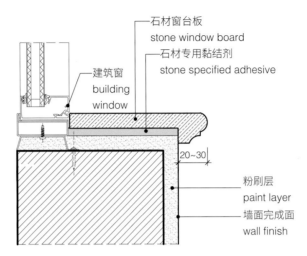

比例 | scale: 1:5

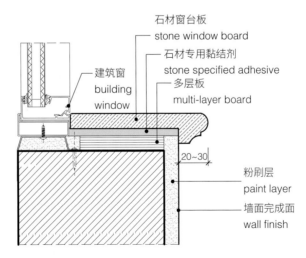

节点图　DETAIL

挂墙式马桶
Wall-mounted Toilet

060P / 061P

重点 / KEY POINTS

注意矮墙的尺寸是否满足入墙式水箱的规格和安装空间要求。大部分入墙式水箱的检修都可通过拆下按钮面板来解决，不需要专门设置检修口。如果考虑加装智能马桶盖板，则需要预留电源和给水点位。

Whether the size of the low wall meets the specifications and installation space of the wall-mounted water tank should be paid more attention. Most of the repairs of wall-mounted water tanks can be solved by removing the button panel without other special fixtures. If intelligent toilet cover is considered to be installed, power supply and water supply point should be reserved.

微信扫码，了解更多关于"挂墙式马桶"的知识

微信扫码，观看"挂墙式马桶"三维节点动图

三维图　PERSPECTIVE

墙面工艺节点　DETAILS OF WALL PROCESSING TECHNIQUES

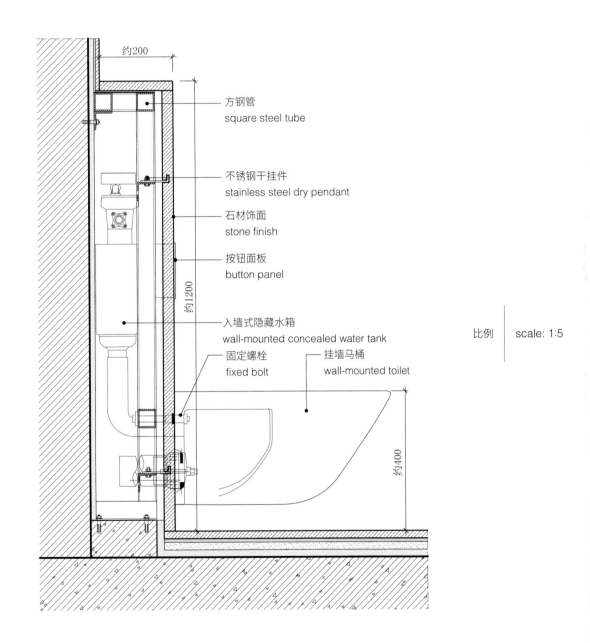

节点图　DETAIL

挂墙式小便器
Wall-mounted Urinal

062P / 063P

重点 / KEY POINTS

冲水感应器有电池供电和电源供电两种，需要根据设计进行选型，做好点位的预留。

There are two kinds of flushing inductors: battery power supply and power supply. The flushing inductor should be selected and the location should be reserved according to the design.

微信扫码，观看"挂墙式小便器"三维节点动图

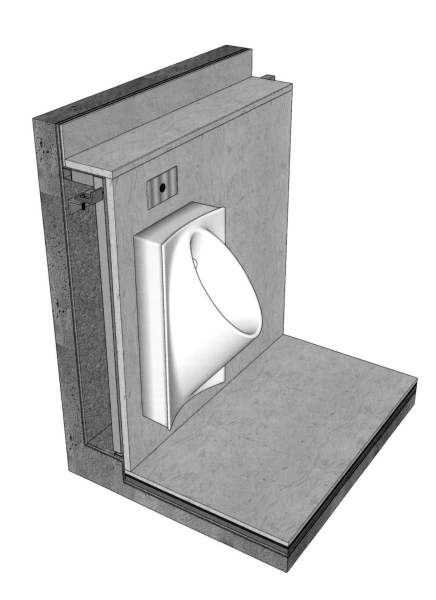

墙面工艺节点　DETAILS OF WALL PROCESSING TECHNIQUES

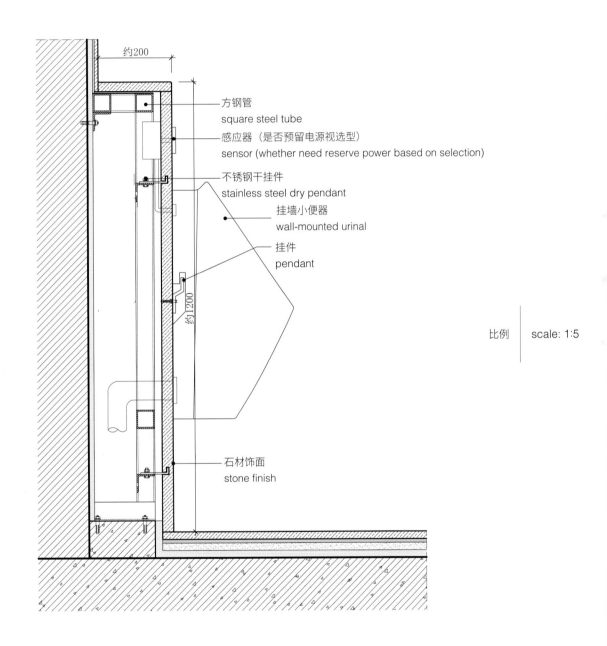

比例　scale: 1:5

节点图　DETAIL

暗水箱蹲便器
Hidden Tank Squatting Toilet

064P / 065P

重点 / KEY POINTS

原理和挂墙式马桶类似，设置蹲便器地坪需要垫高，垫高的尺寸需要和选型结合，自带存水弯的垫高尺寸在 240~300 mm，不带存水弯的垫高尺寸在 150~200 mm。

The principle is similar to wall-mounted toilet. The floor of squatting toilet need to be increased. The size of cushion height should be combined with selected type. The cushion height of contained water storage bend is about 240~300 mm, while the cushion height of non-water storage bend is about 150~200 mm.

微信扫码，观看"暗水箱蹲便器"三维节点动图

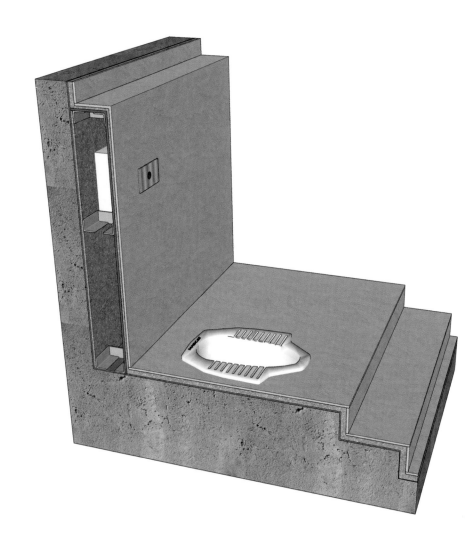

三维图 PERSPECTIVE

墙面工艺节点　DETAILS OF WALL PROCESSING TECHNIQUES

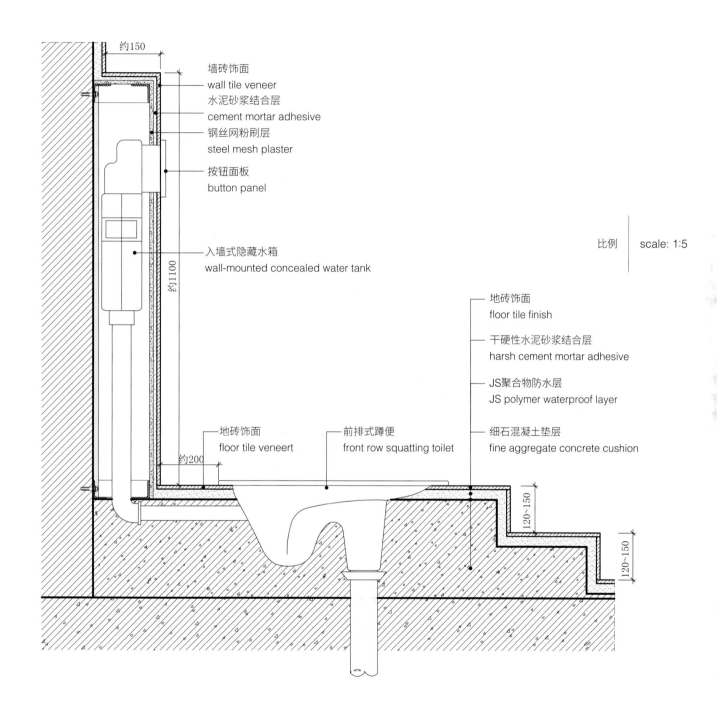

节点图　DETAIL

侧拉式防火卷帘
Side-pull Fire-proof Roll Curtain

066P / 067P

重点 / KEY POINTS

是否设置侧拉式防火卷帘由建筑设计单位决定。

Whether to install side-pull fire-proof roll curtain is decided by the architectural design unit.

微信扫码，观看"侧拉式防火卷帘"三维节点动图

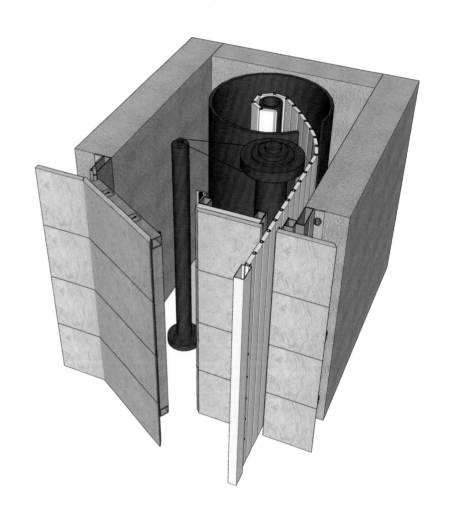

三维图　PERSPECTIVE

墙面工艺节点　DETAILS OF WALL PROCESSING TECHNIQUES

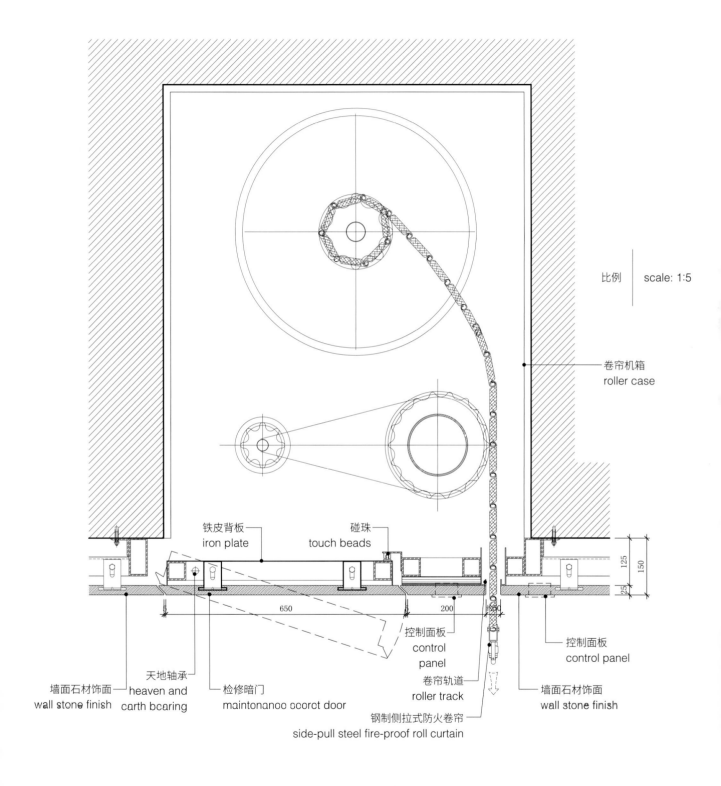

节点图　DETAIL 067

2

吊顶工艺节点
DETAILS OF SUSPENDED CEILING PROCESSING TECHNIQUES

  本篇的主要内容是常用吊顶的工艺做法，涉及的材料有石膏板、金属格栅、木饰面、亚克力、软膜等，涉及的构造有检修口、伸缩缝、挡烟垂壁、消防卷帘等。

  吊顶的造型根据设计方案会有多种表现形式，但是其中的工艺原理是相通的，设计师可以根据这些标准做法来举一反三。在工装设计中，不可避免地会碰到很多和功能及规范有关的构造构件，这也是设计师容易忽略的。本篇提到的一些常见的做法可以供设计师参考，设计师在实际项目中遇到时可以互相印证。

石膏板吊顶　|贴顶式|
Gypsum Board Suspended Ceiling　| Top Stick |

070P / 071P

重点 / KEY POINTS

常用石膏板厚度为 9.5 mm 和 12 mm 两种，天花板一般采用双层 9.5 mm 厚的石膏板，平整度更好，也能有效防止开裂。

Gypsum board of 9.5 mm and 12 mm thick are commonly used, among which the former is employed in ceilings because of its better smoothness and effectiveness in preventing cracking.

微信扫码，观看"石膏板吊顶 |贴顶式|"三维节点动图

三维图　PERSPECTIVE

吊顶工艺节点　DETAILS OF SUSPENDED CEILING PROCESSING TECHNIQUES

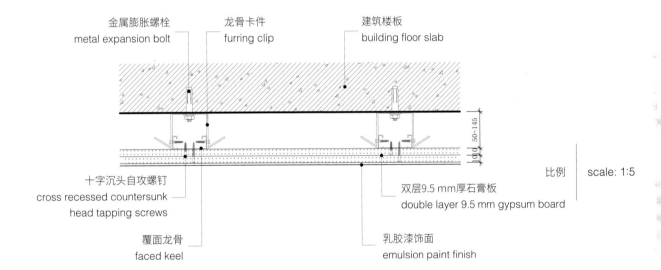

节点图　DETAIL

071

石膏板吊顶　　|悬吊式|
Gypsum Board Suspended Ceiling　　| Suspended |

072P / 073P

重点 / KEY POINTS

天花龙骨分为60系列、50系列、38系列等，根据承载龙骨规格不同命名。装饰工程常用的有50系列、60系列。

According to the specifications of loading bearing keel, ceiling keels are divided into different categories including 60, 50, 38 series, among which 50 and 60 series are commonly used in decoration engineering.

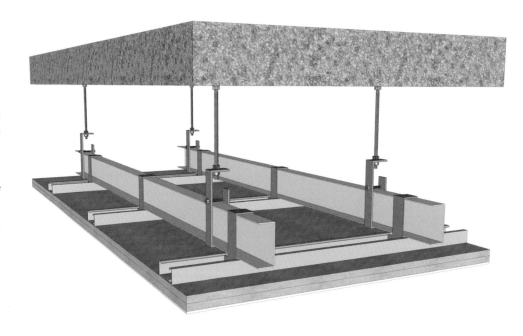

微信扫码，了解更多关于"石膏板吊顶"的知识

微信扫码，观看"石膏板吊顶|悬吊式|"三维节点动图

三维图　PERSPECTIVE

吊顶工艺节点　DETAILS OF SUSPENDED CEILING PROCESSING TECHNIQUES

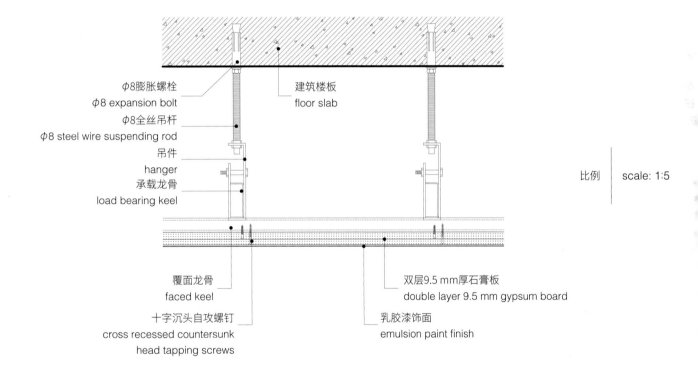

节点图　DETAIL

石膏板吊顶　|卡式承载龙骨|
Gypsum Board Suspended Ceiling　| Clipping Load Bearing Joist |

074P / 075P

重点 / KEY POINTS

卡式承载龙骨即使用有卡齿的可调节 U 形夹来取代主龙骨，直接与覆面龙骨卡合形成骨架体系，适用于安装空间狭小但又需要吊顶的部位。

Cladding joist and adjustable U-clip which owns latch displaces main runner can compose skeleton structure. The structure is suitable for narrow and small place which need suspended ceiling.

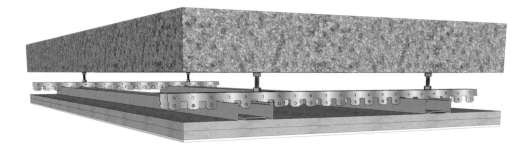

微信扫码，观看"石膏板吊顶|卡式承载龙骨|"三维节点动图

三维图　PERSPECTIVE

吊顶工艺节点　DETAILS OF SUSPENDED CEILING PROCESSING TECHNIQUES

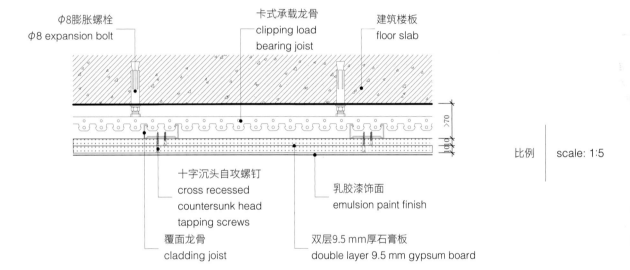

节点图　DETAIL

075

石膏板吊顶　|高低差造型|
Gypsum Board Suspended Ceiling　| High and Low Shape |

076P / 077P

重点 / KEY POINTS

　　高顶和低顶都采用常规轻钢龙骨吊顶形式，连接高低顶的部分一般用扁铁吊装基层板来处理。

Both high and low ceilings apply the form of conventional light steel keel suspended ceilings. Band iron which suspends base board is generally used in the connection portion of the two ceilings.

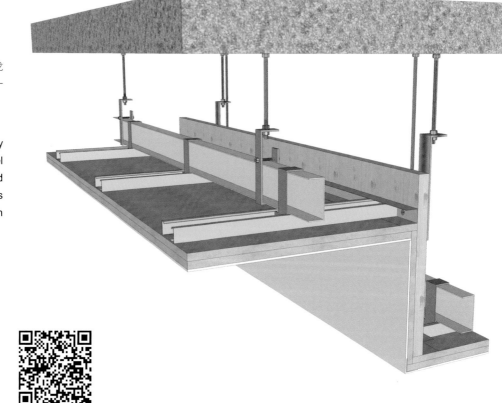

微信扫码，了解更多关于"石膏板吊顶造型"的知识

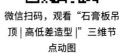

微信扫码，观看"石膏板吊顶|高低差造型|"三维节点动图

三维图　PERSPECTIVE

吊顶工艺节点　DETAILS OF SUSPENDED CEILING PROCESSING TECHNIQUES

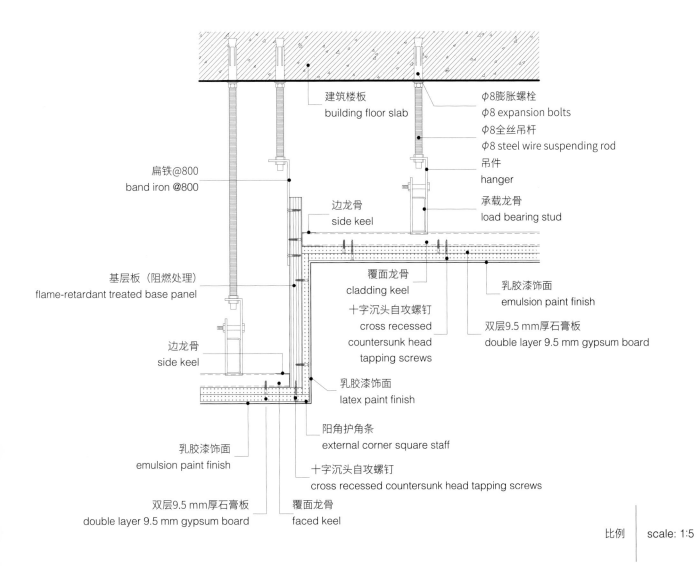

节点图　DETAIL

077

石膏板吊顶　　|阴角石膏线条/顶面石膏线条|
Gypsum Board Suspended Ceiling　　| Inner Corner Line/Top Surface Gypsum Line |

078P / 079P

重点 / KEY POINTS

石膏阴角安装分为粘线法安装工艺和钉线法安装工艺。

The methods of the mounting of gypsum on inner corners include stuck mould and devil float.

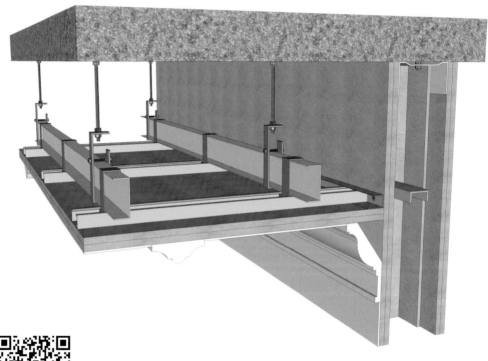

微信扫码，了解更多关于"石膏线条"的知识

微信扫码，观看"石膏板吊顶 | 阴角石线条 / 顶面石膏线条 |"三维节点动图

078　　　　三维图　PERSPECTIVE

吊顶工艺节点　DETAILS OF SUSPENDED CEILING PROCESSING TECHNIQUES

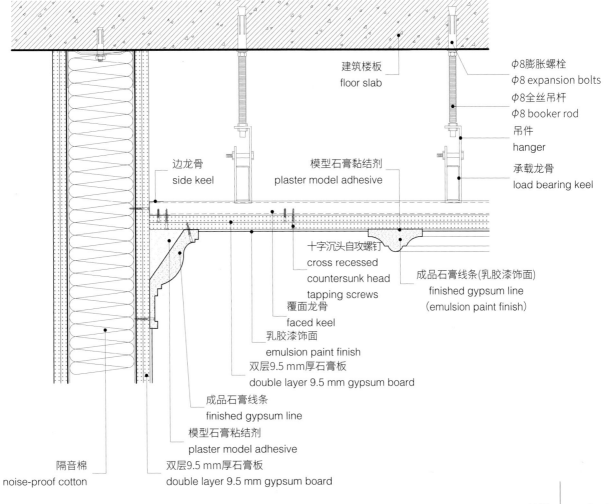

scale: 1:5

节点图　DETAIL

石膏板吊顶　|常规灯槽造型|
Gypsum Board Suspended Ceiling　| Conventional Light Trough Shape |

080P / 081P

重点 / KEY POINTS

灯槽深度及高度不宜过小，否则对出光效果会有影响；灯槽造型的基层板需要注意防火处理，灯槽内安装白铁皮可以起到防火效果，同时也便于灯具安装。

Light trough shouldn't be too shadow or low in order to maintain proper lighting effects. Base panels of light troughs shall be fire retardant. Tinned sheet iron in light troughs can function as fireproof material and ease the installment of lights.

微信扫码，观看"石膏板吊顶 | 常规灯槽造型 |"三维节点动图

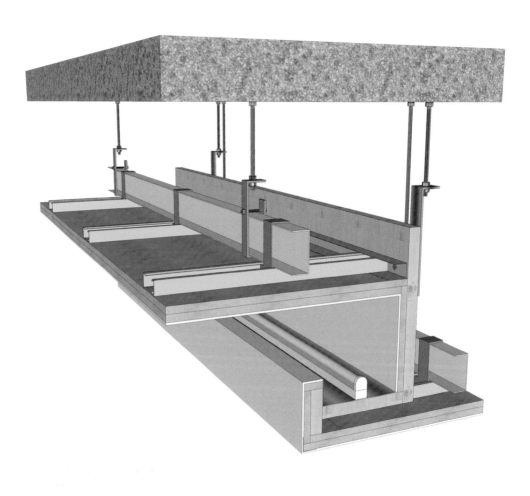

三维图　PERSPECTIVE

吊顶工艺节点 DETAILS OF SUSPENDED CEILING PROCESSING TECHNIQUES

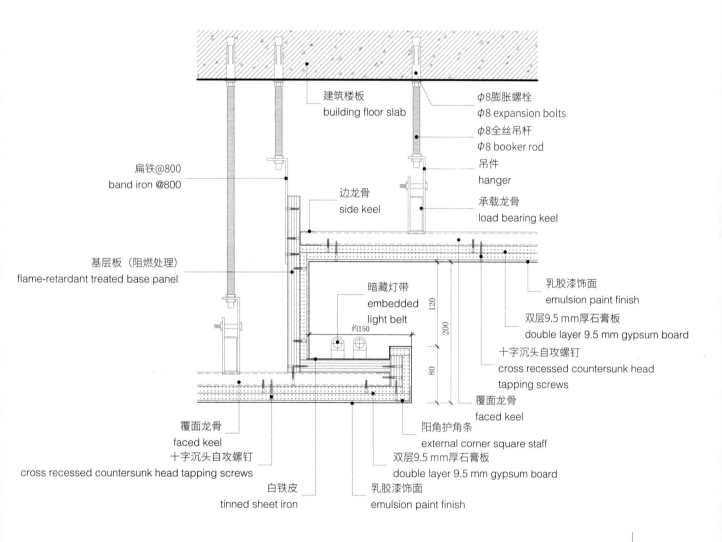

节点图 DETAIL

scale: 1:5

石膏板吊顶　|带石膏线灯槽造型|
Gypsum Board Suspended Ceiling　| Light Trough Shape With Gypsum Line |

082P / 083P

重点 / KEY POINTS

　　灯槽内灯具常规选用 T5 灯管或 LED 灯带，如果选用后者则需注意 LED 灯带自带变压器的摆放位置。

Conventionally, lights in troughs are T5 lamp tube or LED lamp belt. In the use of the latter, the location of their transformers should be paid attention to.

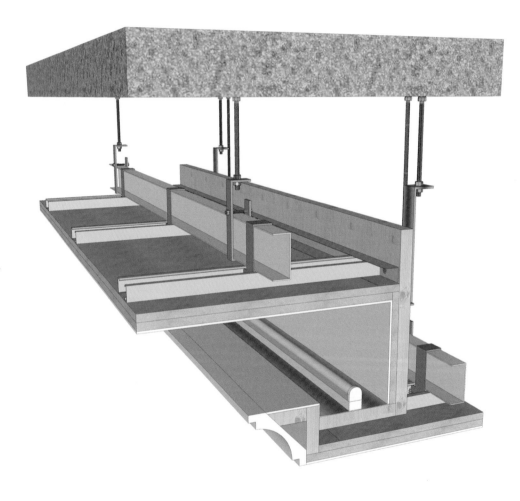

微信扫码，观看"石膏板吊顶 | 带石膏线灯槽造型 |"三维节点动图

三维图　PERSPECTIVE

吊顶工艺节点　DETAILS OF SUSPENDED CEILING PROCESSING TECHNIQUES

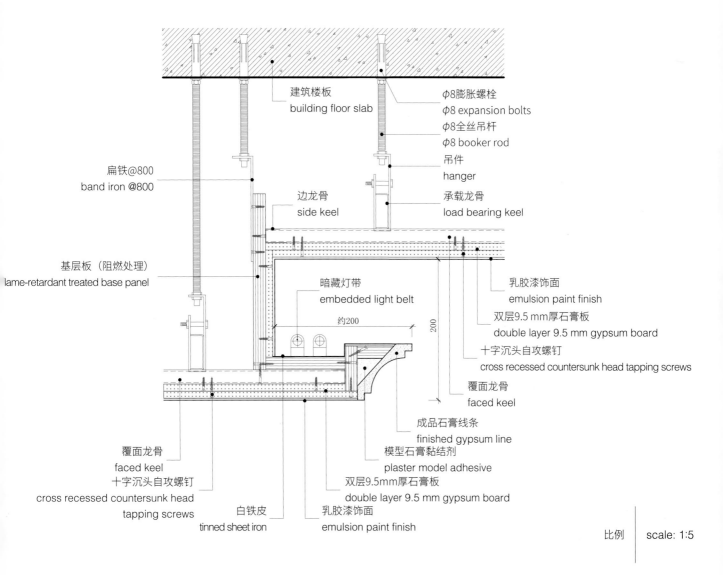

节点图　DETAIL

石膏板吊顶　|弧形石膏线灯槽造型|
Gypsum Board Suspended Ceiling　|Light Trough Shape with Compass Gypsum Line|

084P / 084P

重点 / KEY POINTS

弧形造型一般可采用石膏板弯曲或 GRG（加强纤维石膏板）成品两种方式，前者现场施工即可，后者需厂家定制现场安装，但是复杂造型后者效果更佳。

Cove shape can generally be achieved with bent gypsum board or finished GRG. While bent gypsum boards can be made on site, GRG require being made to order and set on site on the spot yet can better fit complex shape.

微信扫码，观看"石膏板吊顶 | 弧形石膏线灯槽造型 |"三维节点动图

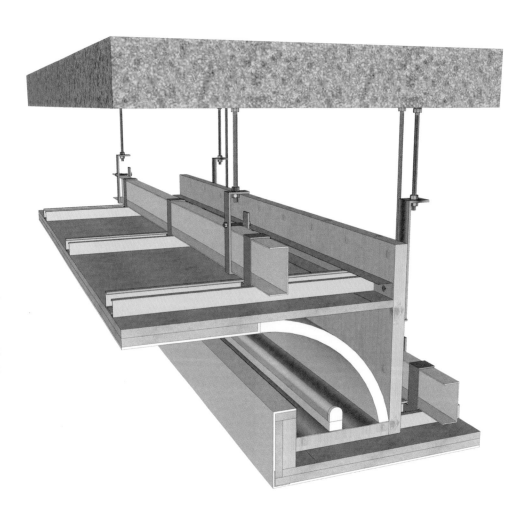

三维图　PERSPECTIVE

吊顶工艺节点　DETAILS OF SUSPENDED CEILING PROCESSING TECHNIQUES

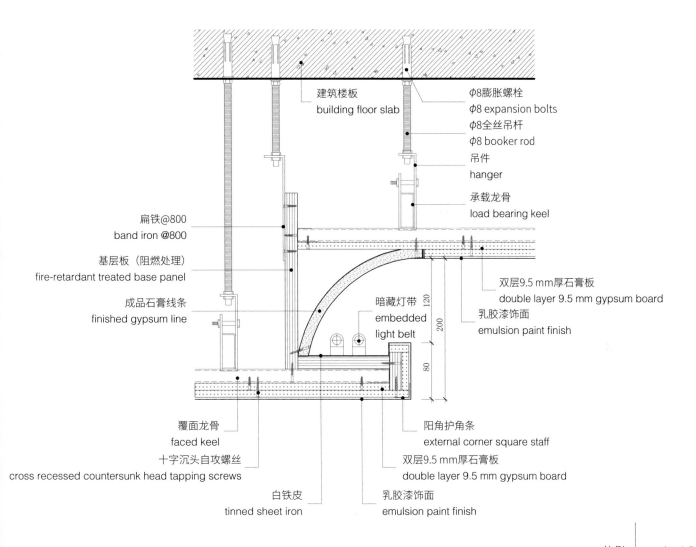

节点图　DETAIL

石膏板吊顶　　|靠墙风口带灯槽造型|
Gypsum Board Suspended Ceiling　　| Light Trough Shape with Wall Whirl Tube |

086P / 087P

重点 / KEY POINTS

灯槽距离墙面尺寸不宜小于150 mm，否则会影响送回风的效果。

Distance between light troughs and walls shouldn't be less than 150 mm to keep air supply effective.

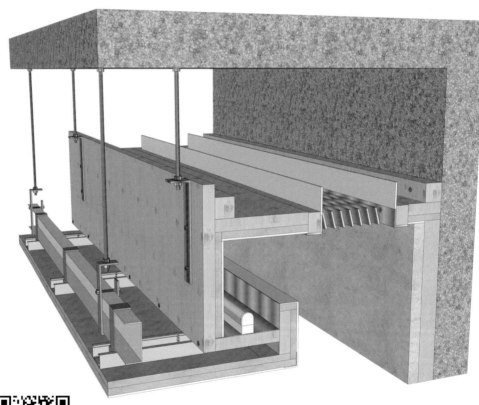

微信扫码，了解更多关于"装饰风口"的知识

微信扫码，观看"石膏板吊顶|靠墙风口带灯槽造型|"三维节点动图

吊顶工艺节点　DETAILS OF SUSPENDED CEILING PROCESSING TECHNIQUES

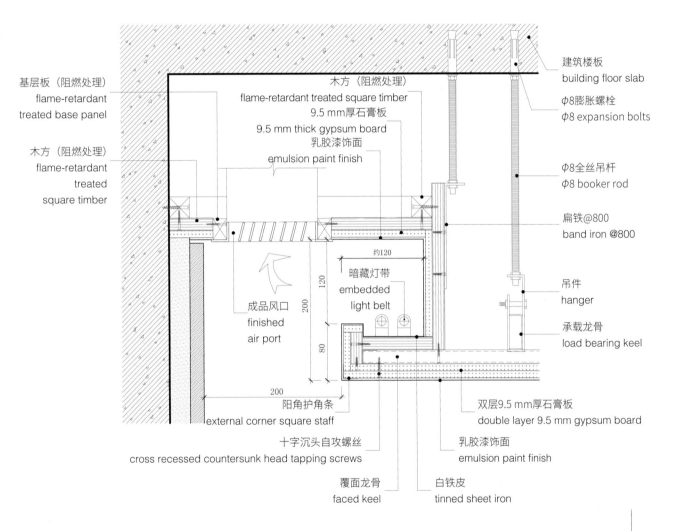

节点图　DETAIL

石膏板吊顶　|灯槽带风口造型|
Gypsum Board Suspended Ceiling　| Light Trough Shape with Whirl Tube |

088P / 089P

重点 / KEY POINTS

安装风口的位置，应先用木方或木条进行加固，便于后期成品风口的安装固定。

The place for air ports should be reinforced by square wood and wood strips, preparing for later installment and fixing of finished air pots.

微信扫码，观看"石膏板吊顶 | 灯槽带风口造型 |"
三维节点动图

三维图　PERSPECTIVE

吊顶工艺节点　DETAILS OF SUSPENDED CEILING PROCESSING TECHNIQUES

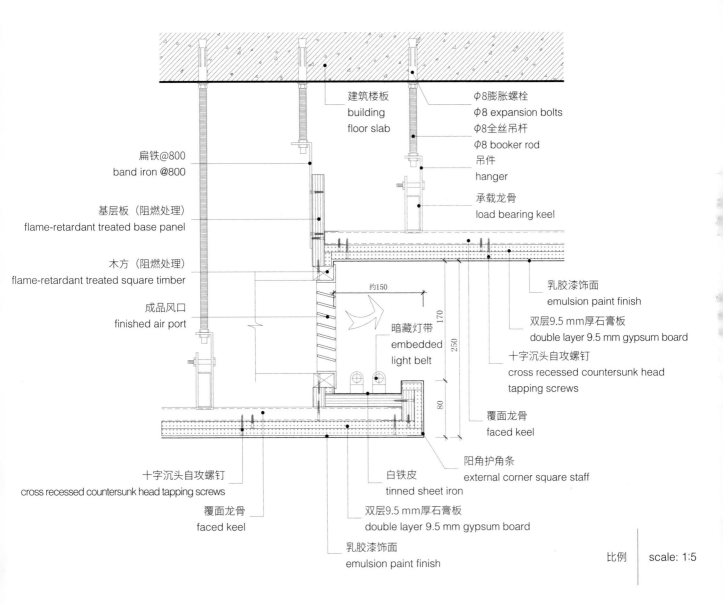

节点图　DETAIL

吊顶　　|顶面墙角留缝造型|
Suspended Ceiling　　| Top Surface Corner with Chute |

090P / 091P

重点 / KEY POINTS

顶角留槽的做法对工艺要求较高，同时设计时需注意留槽造型尽量不要跨越不同高差，否则难以跟通。

The construction of top surface with chute have a very high demand, meanwhile it should be carefully designed instead of crossing step. Otherwise, the chute will not be straight.

微信扫码，观看"吊顶 |
顶面墙角留缝造型 |"
三维节点动图

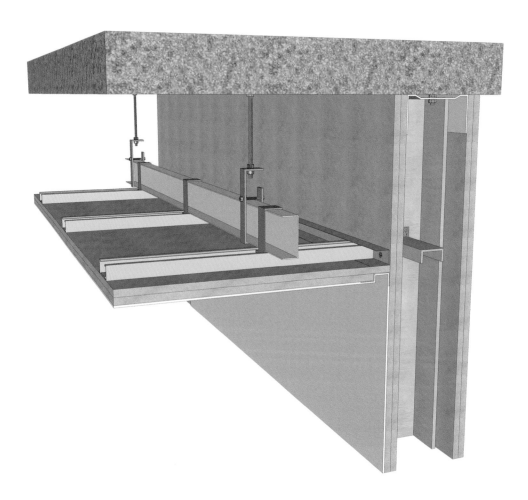

三维图　PERSPECTIVE

吊顶工艺节点　DETAILS OF SUSPENDED CEILING PROCESSING TECHNIQUES

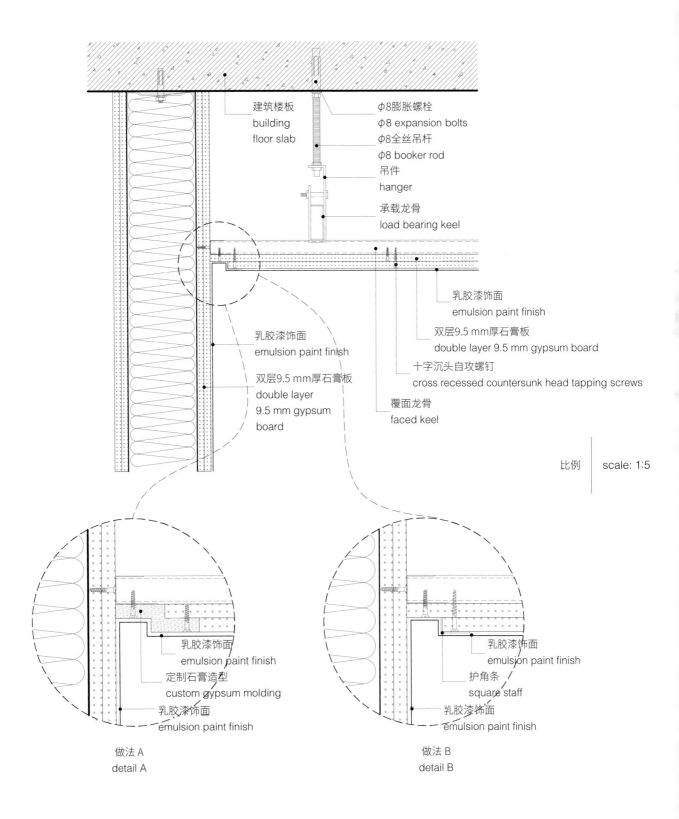

节点图　DETAIL

石膏板吊顶　|顶面留缝造型|
Gypsum Board Suspended Ceiling　| Top Surface with Chute |

092P / 093P

重点 / KEY POINTS

　　石膏板棚留缝的常见宽度尺寸有 10 mm，15 mm，20 mm，高度以一块或两块石膏板厚为好，10~20 mm。

The common width of top surface 's chute is generally 10 mm, 15 mm, 20 mm wide and properly 10~20 mm high, as the thickness of one or two gypsum boards.

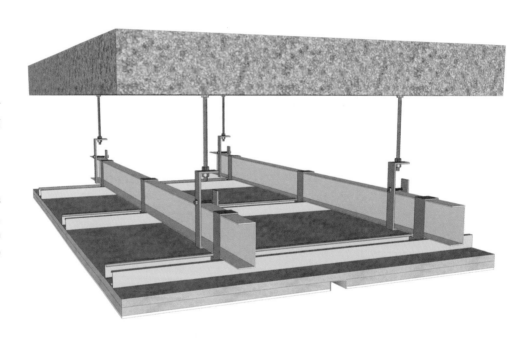

微信扫码，观看"石膏板吊顶 | 顶面留缝造型 |"
三维节点动图

三维图　PERSPECTIVE

吊顶工艺节点　DETAILS OF SUSPENDED CEILING PROCESSING TECHNIQUES

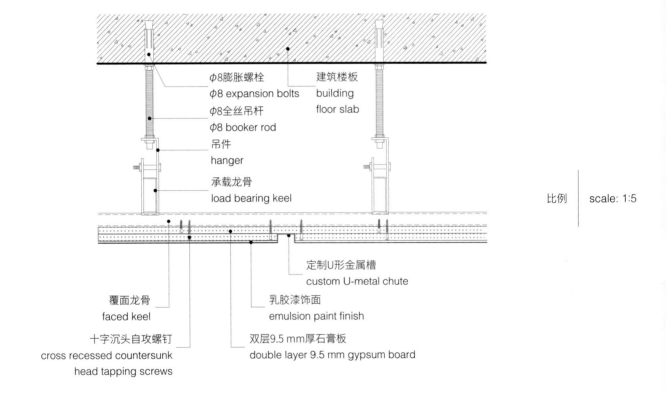

比例　scale: 1:5

节点图　DETAIL

明装式窗帘盒天花
Open Equipping Curtain Box

094P / 095P

重点 / KEY POINTS

当吊顶高度不允许或设计风格需要时可采用明装式窗帘盒。窗帘盒宽度一般为 200 mm（双轨），若是单轨，可以考虑 150 mm 宽度。

When there's no high enough suspended ceiling or specific needs for style, open equipping curtain box can be put into use. Width is commonly around 200 mm (when it's in the situation of two-rail) wide, while a width of 150 mm maybe proper for monorail.

微信扫码，了解更多关于"窗帘盒"的知识

微信扫码，观看"明装式窗帘盒天花"三维节点动图

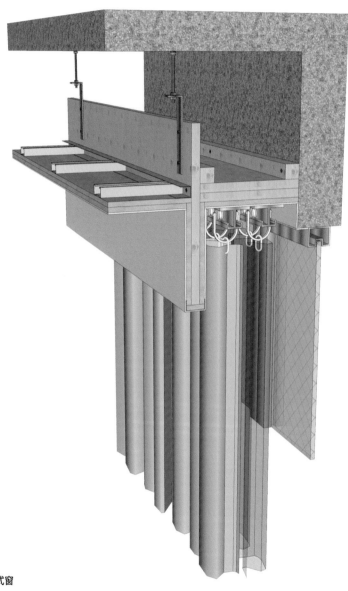

三维图　PERSPECTIVE

吊顶工艺节点　DETAILS OF SUSPENDED CEILING PROCESSING TECHNIQUES

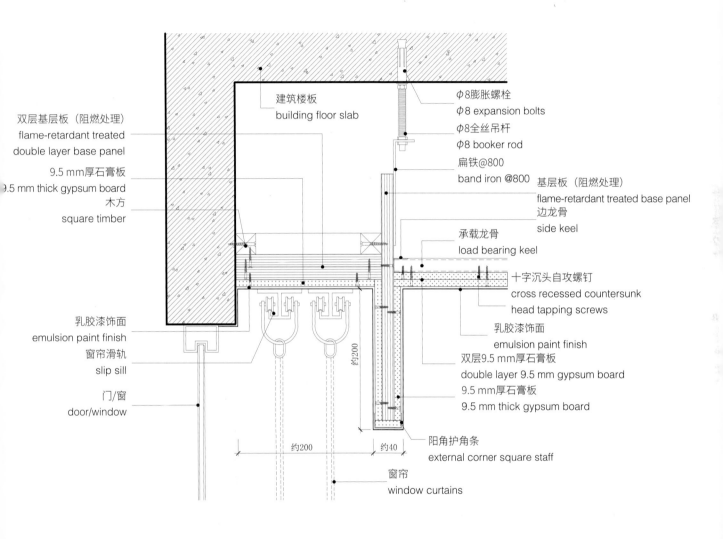

节点图　DETAIL

暗装式窗帘盒天花
Submerged Curtain Box

096P / 097P

重点 / KEY POINTS

窗帘盒宽度一般为 200 mm（双帘），若是单帘，可以考虑 150 mm。双帘指一层纱帘一层遮光帘，单帘就是一层遮光帘。

Width of double-curtain is generally 200 mm wide, and a width of 150 mm can be considered when it comes to single-curtain. Double-curtain refers to a layer of lace curtain and window blind, while single-curtain only consists a layer of window blind.

微信扫码，观看"暗装式窗帘盒天花"三维节点动图

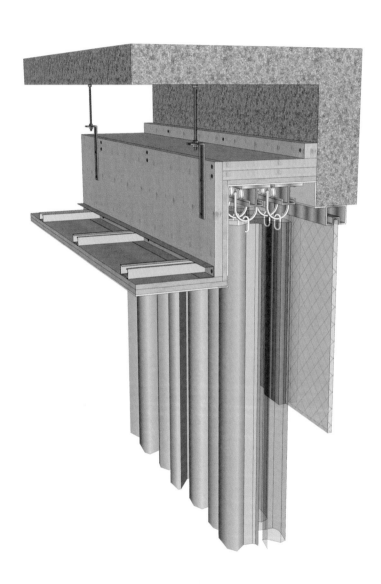

三维图 PERSPECTIVE

吊顶工艺节点 DETAILS OF SUSPENDED CEILING PROCESSING TECHNIQUES

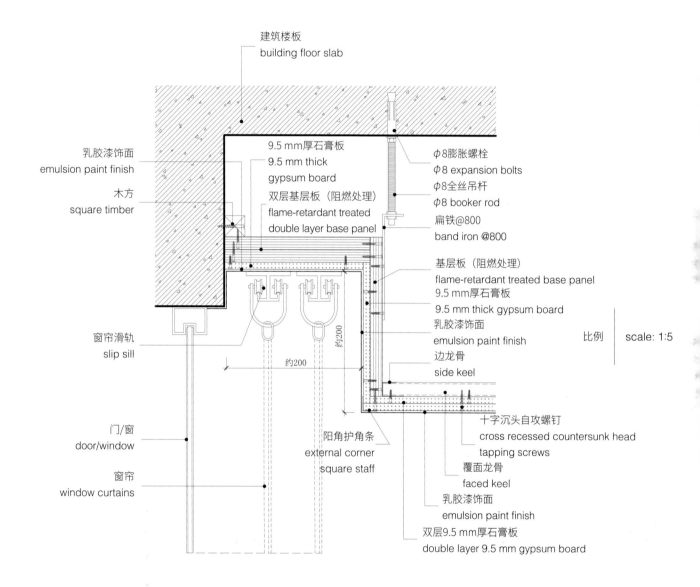

节点图 DETAIL

明装式窗帘盒天花　|低于窗户|
Open Equipping Curtain Box　|Below the Window|

重点 / KEY POINTS

窗帘盒位置低于窗户时需要做相应的竖向挡板和窗框相接，并对挡板朝向窗户一侧进行饰面处理。这一原理也适用于玻璃幕墙。

When the position of the curtain box is lower than the window, the corresponding vertical baffle should be connected with the window frame, and the window side of the baffle should be treated. This principle also applies to glass curtain wall.

微信扫码，观看"明装式窗帘盒天花 | 低于窗户 |"
三维节点动图

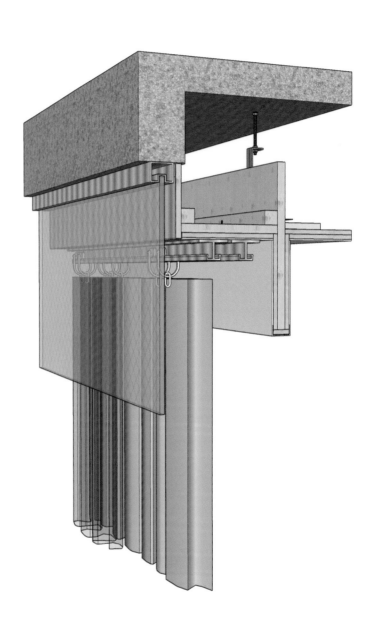

三维图　PERSPECTIVE

吊顶工艺节点　DETAILS OF SUSPENDED CEILING PROCESSING TECHNIQUES

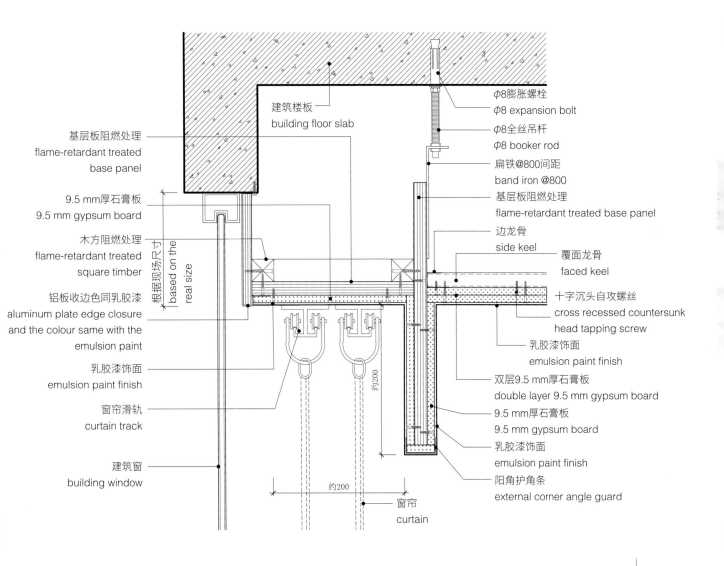

节点图　DETAIL

暗装式窗帘盒天花　|低于窗户|
Submerged Curtain Box　|Below the Window|

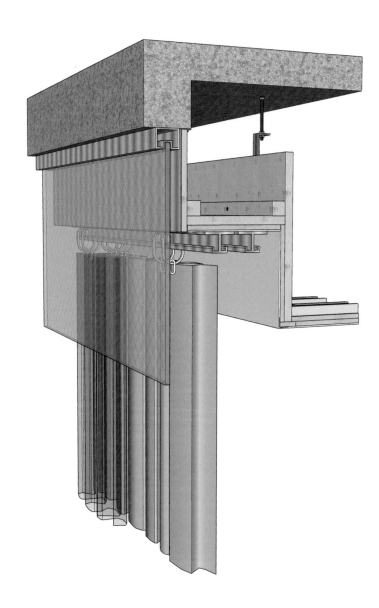

微信扫码，观看"暗装式窗帘盒天花 | 低于窗户 |"三维节点动图

三维图　PERSPECTIVE

吊顶工艺节点　DETAILS OF SUSPENDED CEILING PROCESSING TECHNIQUES

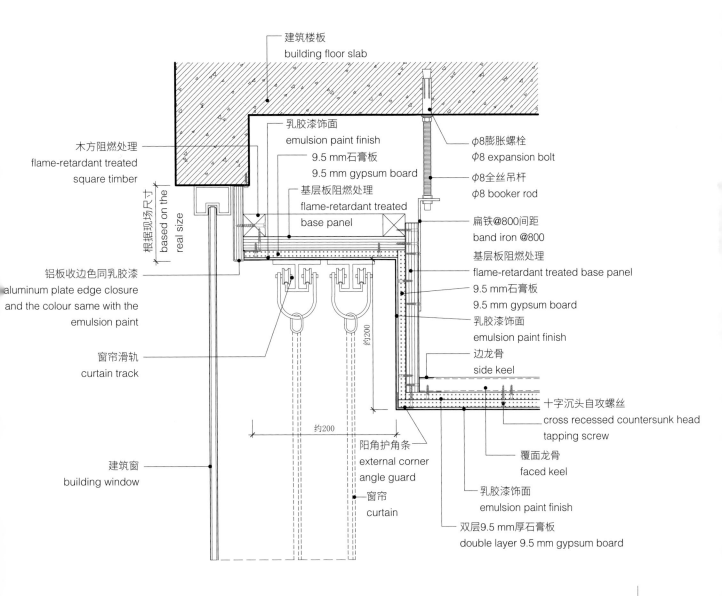

节点图　DETAIL

暗装式投影幕布　　|靠墙|
Submerged Projection Screen　|Against the wall|

102P / 103P

重点 / KEY POINTS

需要根据幕布的规格尺寸进行天花造型的设计，如果采用大型幕布需要考虑相应的构造承重。

The ceiling is designed based on the size of the projection screen. If a large curtain is used, the corresponding structural load-bearing should be taken into account.

微信扫码，观看"暗装式投影幕布 | 靠墙 |"三维节点动图

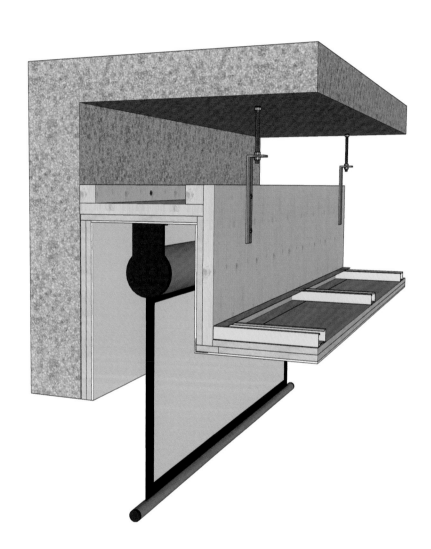

吊顶工艺节点　DETAILS OF SUSPENDED CEILING PROCESSING TECHNIQUES

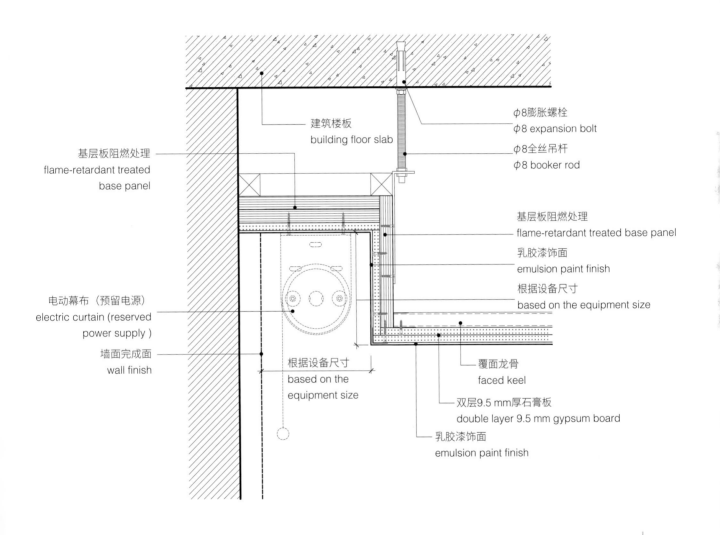

节点图　DETAIL

暗装式投影幕布　|居中|
Submerged Projection Screen　|Center|

104P / 105P

重点 / KEY POINTS

投影幕设置在空间的中间位置时，可以考虑这种处理方式。

When the projection screen is located in the middle of the space, this method can be considered.

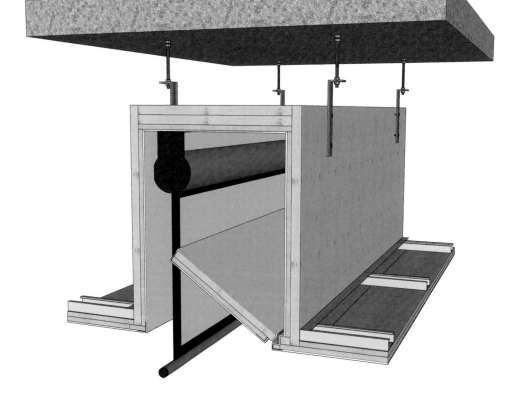

微信扫码，观看"暗装式投影幕布 | 居中 |"三维节点动图

吊顶工艺节点　DETAILS OF SUSPENDED CEILING PROCESSING TECHNIQUES

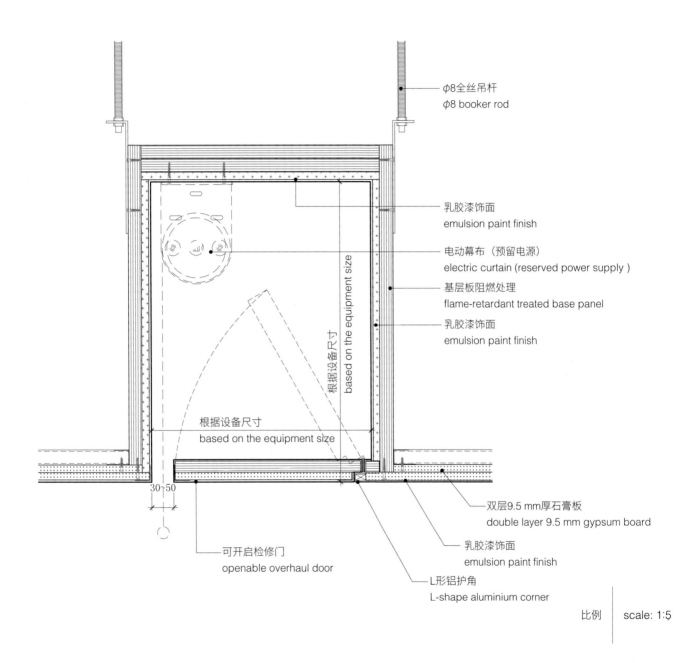

节点图　DETAIL

升降投影仪
Lift projector

106P / 107P

重点 / KEY POINTS

根据投影仪规格来选择相应的升降设备。注意当投影仪处于降下状态时,投影路径上不要有遮挡物。

Selecting the lifting equipment should consider the specifications of the projector. In addition, if the projector is falling, there should be no obstruction on the projection path.

微信扫码,观看"升降投影仪"三维节点动图

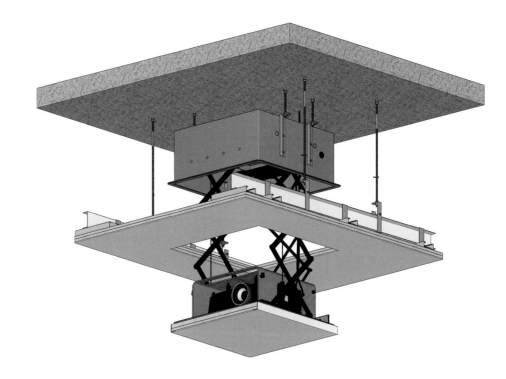

吊顶工艺节点　DETAILS OF SUSPENDED CEILING PROCESSING TECHNIQUES

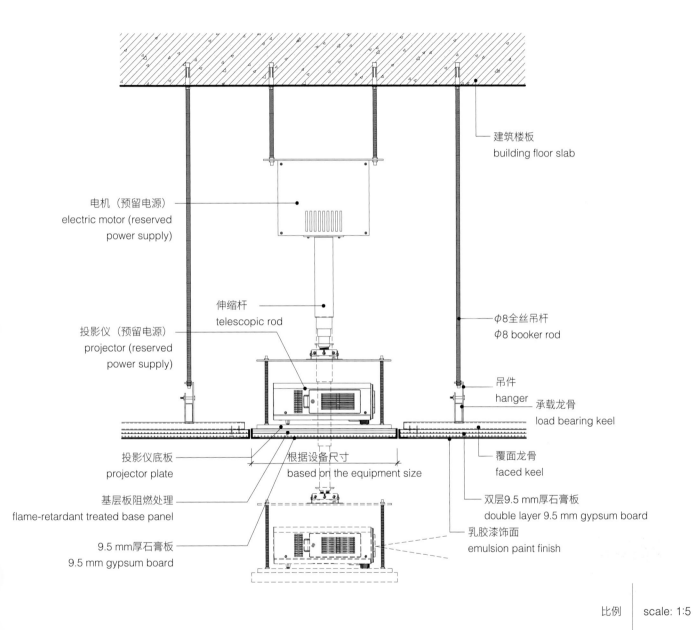

节点图　DETAIL

嵌入式顶花洒
Embedded Ceiling Sprinkler

108P / 109P

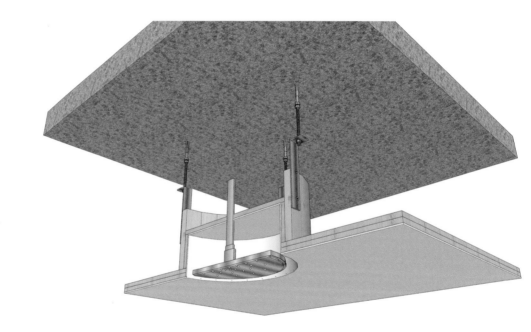

重点 / KEY POINTS

对花洒的规格和安装方式要有了解，预留好花洒和天花造型之间的操作空间。

More understanding is needed on the specifications and installation methods of sprinkler, and the operation space between sprinkler and ceiling modeling should be reserved.

微信扫码，了解更多关于"顶喷花洒"的知识

微信扫码，观看"嵌入式顶花洒"三维节点动图

三维图　PERSPECTIVE

吊顶工艺节点　DETAILS OF SUSPENDED CEILING PROCESSING TECHNIQUES

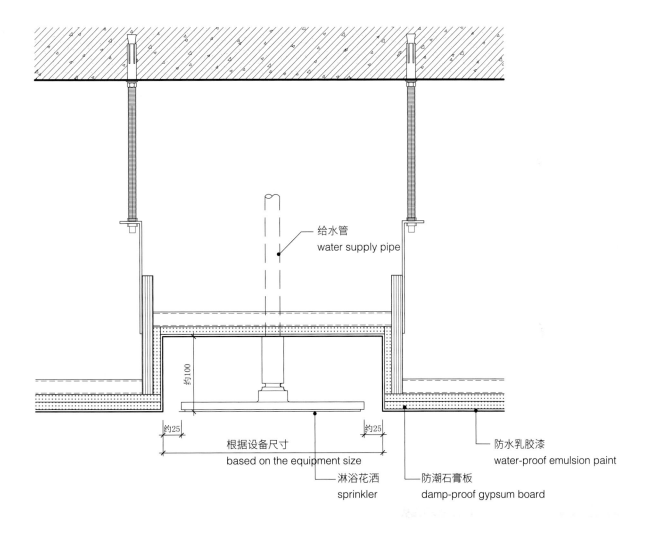

节点图　DETAIL

涂料顶面与涂料墙面交接天花
The Connection of Coating Ceiling and Coating Wall

110P / 111P

重点 / KEY POINTS

天花石膏板与墙面石膏板墙交接需要安装边龙骨。

Side keel is needed in the connection portion of gypsum ceiling and wall.

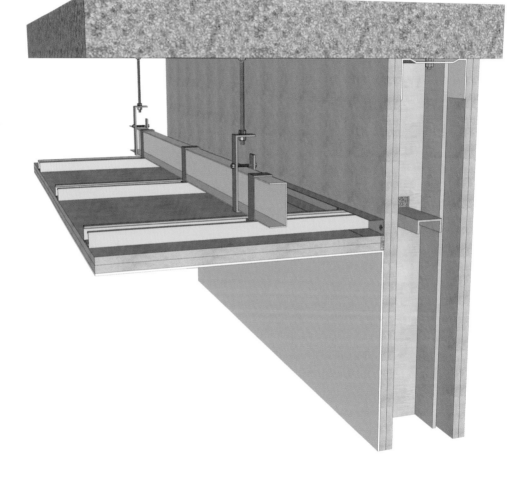

微信扫码，观看"涂料顶面与涂料墙面交接天花"三维节点动图

吊顶工艺节点　DETAILS OF SUSPENDED CEILING PROCESSING TECHNIQUES

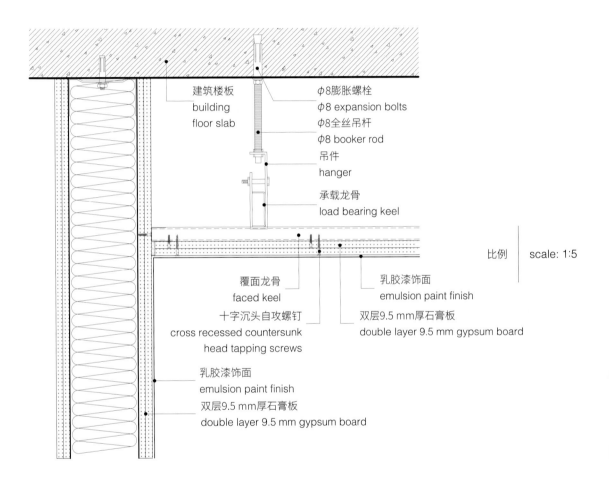

比例　scale: 1:5

节点图　DETAIL

涂料顶面与石材墙面交接天花
The Connection of Coating Ceiling and Stone Furnishing Wall

112P / 113P

重点 / KEY POINTS

施工顺序为：先做棚面吊顶龙骨，再做墙面石材。

The construction sequence is from suspended ceiling to stone furnishing wall.

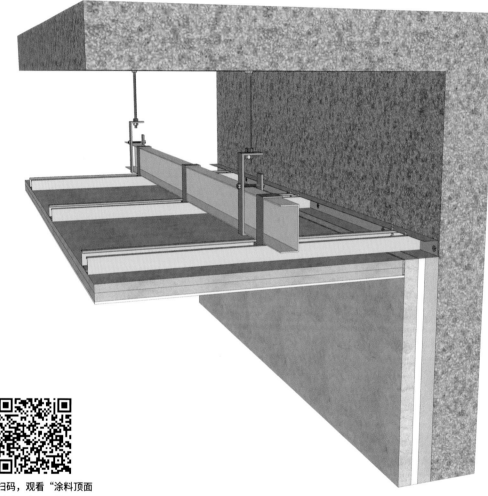

微信扫码，了解更多关于"天花和墙面收口"的知识

微信扫码，观看"涂料顶面与石材墙面交接天花"三维节点动图

三维图 PERSPECTIVE

吊顶工艺节点　DETAILS OF SUSPENDED CEILING PROCESSING TECHNIQUES

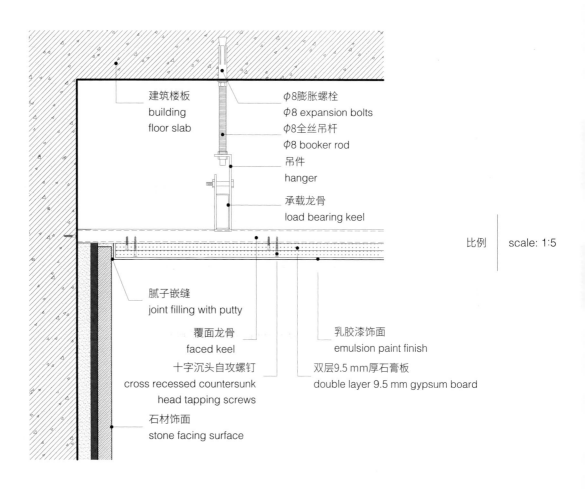

比例　scale: 1:5

节点图　DETAIL

113

矩形金属格栅天花
Metal Grille Ceiling

114P / 115P

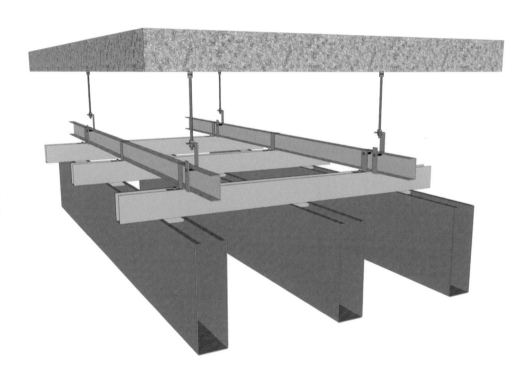

重点 / KEY POINTS

金属格栅的规格尺寸可以定制，定制要考虑成本问题。

The size of metal grids/grills can be customized, yet the cost is to be considered.

微信扫码，了解更多关于"金属天花"的知识

微信扫码，观看"矩形金属格栅天花"三维节点动图

吊顶工艺节点　DETAILS OF SUSPENDED CEILING PROCESSING TECHNIQUES

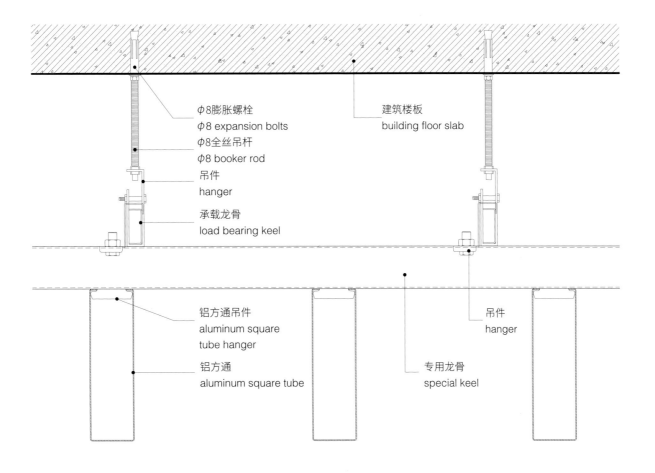

节点图　DETAIL

115

圆形金属格栅天花
Circular Metal Grille Ceiling

116P / 117P

重点 / KEY POINTS

在采用了金属格栅天花的情况下，需要考虑如何处理暴露的建筑楼板，灯具与格栅的安装方式也要考虑清楚。

When circular metal grille ceiling is used, the disposition of exposed floor slabs and the installment measure of lamp and grille require consideration.

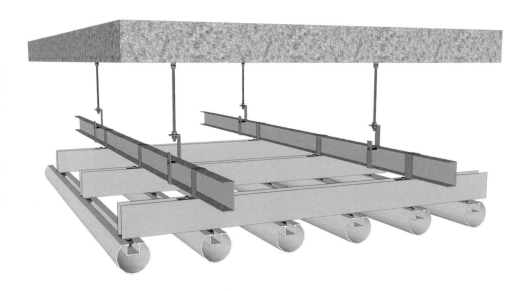

微信扫码，观看"圆形金属格栅天花"三维节点动图

吊顶工艺节点　DETAILS OF SUSPENDED CEILING PROCESSING TECHNIQUES

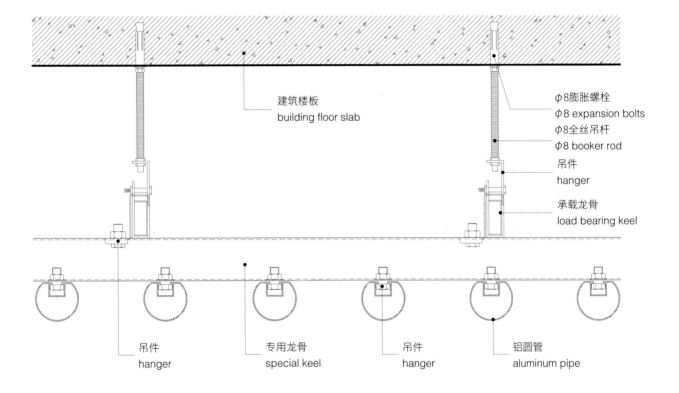

比例　scale: 1:5

节点图　DETAIL

木饰面吊顶天花
Wood Facing Surface Suspended Ceiling

118P / 119P

重点 / KEY POINTS

在防火等级要求高的项目中，不能在天花上使用木饰面，可以用木纹转印铝板或者复合木饰面（木皮覆于金属板或石膏板基层上）来替代。

In projects highly demanding fire retardation, in place of the wood furnish, graining aluminum plates and composite wood furnishing(i.e. metal or gypsum boards with veneer covering the base) should be used on ceilings.

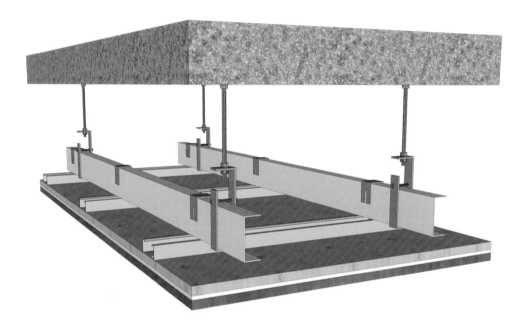

微信扫码，了解更多关于"木饰面天花"的知识

微信扫码，观看"木饰面吊顶天花"三维节点动图

三维图　PERSPECTIVE

吊顶工艺节点　DETAILS OF SUSPENDED CEILING PROCESSING TECHNIQUES

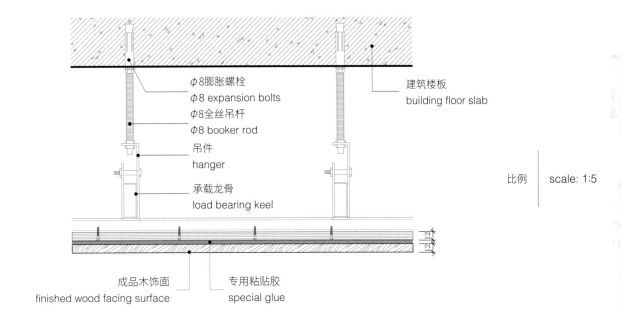

节点图　DETAIL

软膜吊顶天花
Stretch Ceiling Suspended Ceiling

120P / 121P

重点 / KEY POINTS

软膜天花内一般也需腻子批嵌后刷乳胶漆，或是贴上白铁皮。软膜天花龙骨分为扁码（H码）、F码、双扣码。在有防火要求的时候应该选用A级膜，但是A级膜的尺寸规格会比常规软膜的要小。

Stretch ceiling keel always needs joint filling with putty. Afterward, it also need to be brushed elusion paint or pasted with tinned sheet iron. Stretch ceiling keel is classified into flat size(size H), size F and double-clip size . Film A should be used to meet fireproof requirements, but its format dimension is smaller than conventional stretch ceiling.

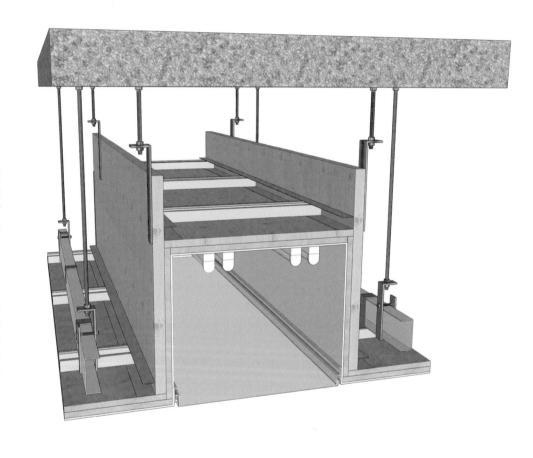

微信扫码，了解更多关于"软膜天花"的知识

微信扫码，观看"软膜吊顶天花"三维节点动图

吊顶工艺节点　DETAILS OF SUSPENDED CEILING PROCESSING TECHNIQUES

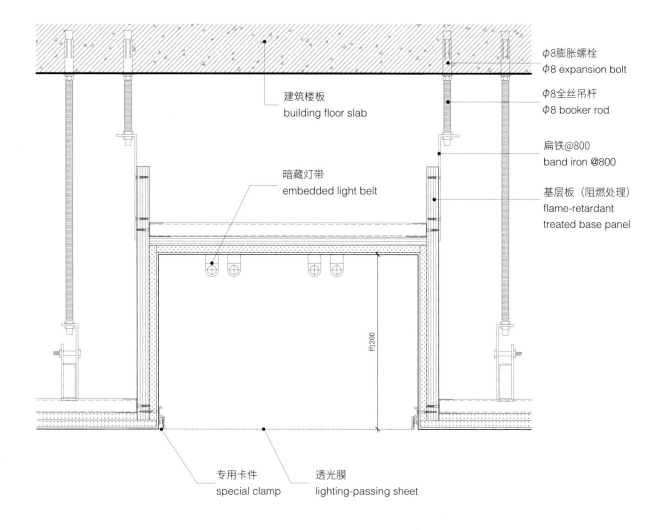

节点图　DETAIL

亚克力吊顶天花
Acrylic Suspended Ceiling

122P / 123P

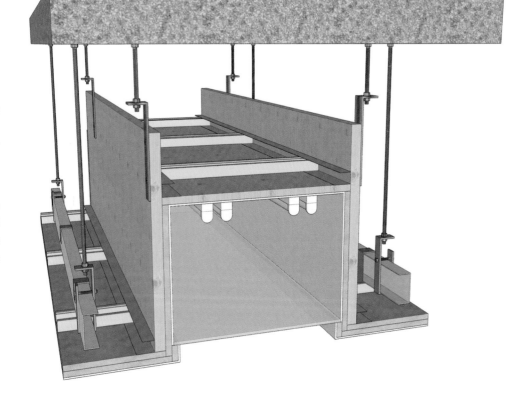

重点 / KEY POINTS

亚克力俗称有机玻璃，在装饰设计中经常被用作透光板、灯箱片等，装饰设计中常用的亚克力板材厚度在 1~20 mm 之间。

Acrylic, known as Perspex, is regularly used as light-passing board and lamp house pieces and has a width of 1~20 mm in decoration engineering.

微信扫码，观看"亚克力吊顶天花"三维节点动图

吊顶工艺节点　DETAILS OF SUSPENDED CEILING PROCESSING TECHNIQUES

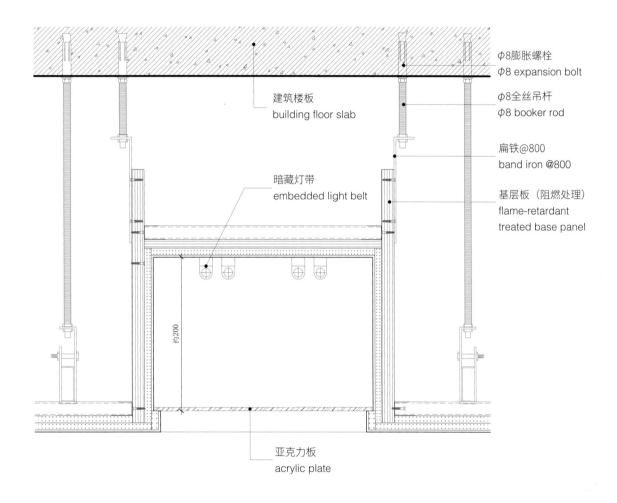

节点图　DETAIL

伸缩缝工艺天花
Control Joint Ceiling

124P / 125P

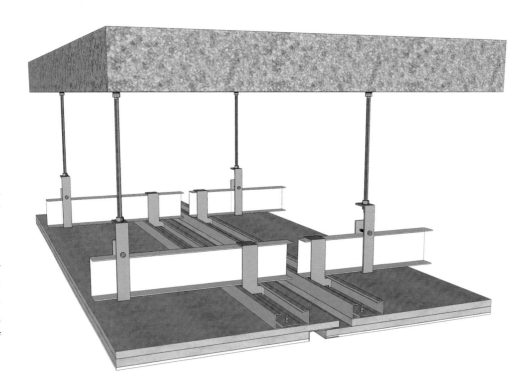

重点 / KEY POINTS

当石膏板天花吊顶面积过大或过长时，为避免由于变形或伸缩导致天花变形开裂而设置软连接结构，让天花在伸缩时能有余地缓冲。

When control joint ceiling is overly large or long, flexible connections can be employed to offer enough room for ceilings' expansion, thus ceilings won't crack as a consequence of stretching or deformation.

微信扫码，了解更多关于"变形缝装饰处理"的知识

微信扫码，观看"伸缩缝工艺天花"三维节点动图

吊顶工艺节点　DETAILS OF SUSPENDED CEILING PROCESSING TECHNIQUES

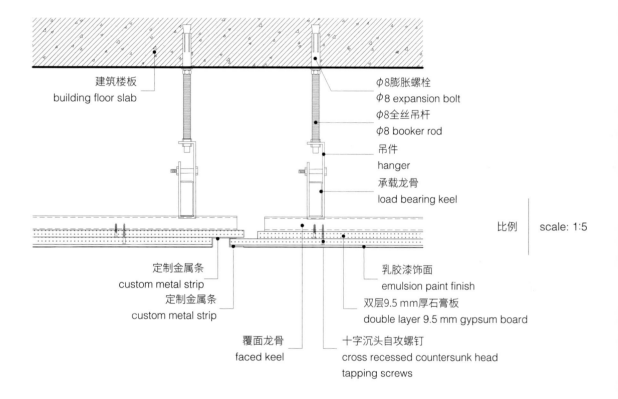

节点图　DETAIL

成品检修口天花
Inspection Opening Suspended Ceiling

126P / 127P

重点 / KEY POINTS

检修口的大小可以根据需求定制，一般不上人检修口为300 mm×300 mm左右；上人检修口为450 mm×450 mm左右。

The size of inspection opening can be customized as needed. Generally, sight hole is around 300 mm×300 mm large, manhole 450 mm×450 mm.

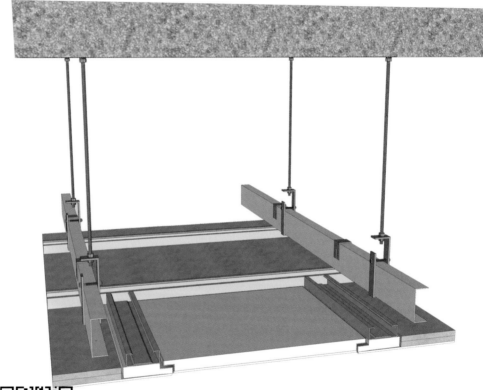

微信扫码，了解更多关于"检修口"的知识

微信扫码，观看"成品检修口天花"三维节点动图

三维图　PERSPECTIVE

吊顶工艺节点　DETAILS OF SUSPENDED CEILING PROCESSING TECHNIQUES

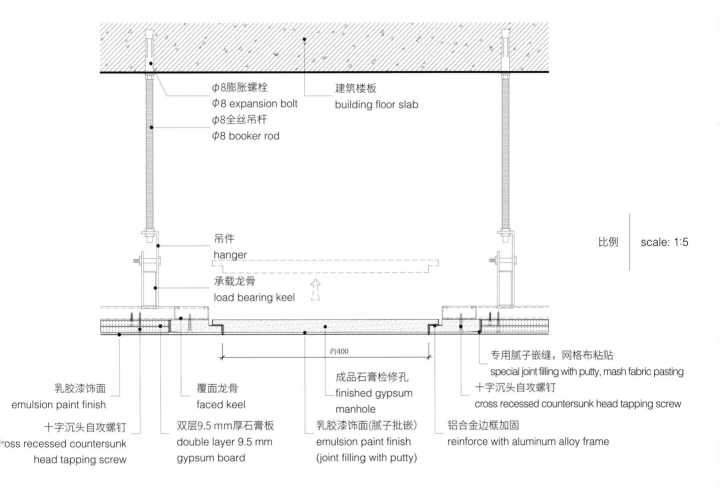

节点图　DETAIL

反支撑工艺天花一
Counteracting Bearing Ceiling 1

128P / 129P

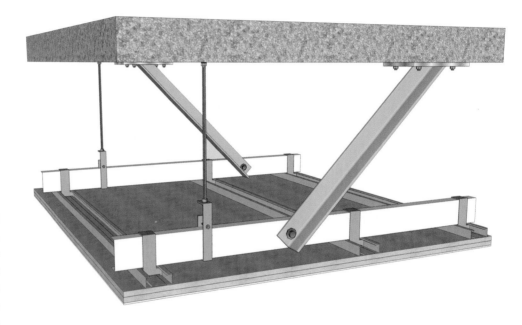

重点 / KEY POINTS

当吊筋长度大于 1 500 mm 时，应设置反支撑，反支撑（镀锌角钢）应与主龙骨连接，角度为 45°，反支撑间距一般为 2 mm 左右。

When the length of hanging steel bar is greater than 1 500 mm, counteracting bearing (i.e. zinc-coated angle steel)should be set up. Countering bearing should be connected to carrying channel. Its angle should be 45°, and its space is commonly about 2 mm.

微信扫码，了解更多关于"反支撑"的知识

微信扫码，观看"反支撑工艺天花一"三维节点动图

吊顶工艺节点　DETAILS OF SUSPENDED CEILING PROCESSING TECHNIQUES

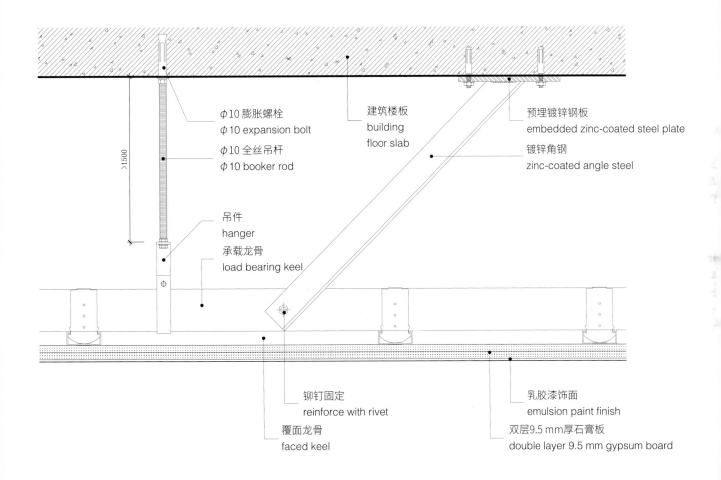

节点图　DETAIL

反支撑工艺天花二
Counteracting Bearing Ceiling 2

130P / 131P

重点 / KEY POINTS

当吊筋大于 1 500 mm 时，需要通过结构工程计算设置，用角钢或主龙骨与楼板、吊顶连接。

If hanging steel bar is longer than 1 500 mm, they should be connected with floor slabs and suspended ceiling with angle steel and carrying channel after structural engineering calculations.

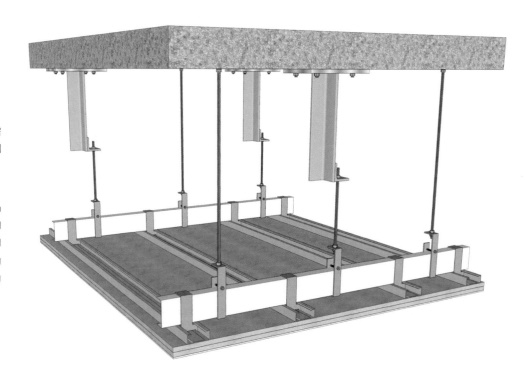

微信扫码，观看"反支撑工艺天花二"三维节点动图

三维图 PERSPECTIVE

吊顶工艺节点　DETAILS OF SUSPENDED CEILING PROCESSING TECHNIQUES

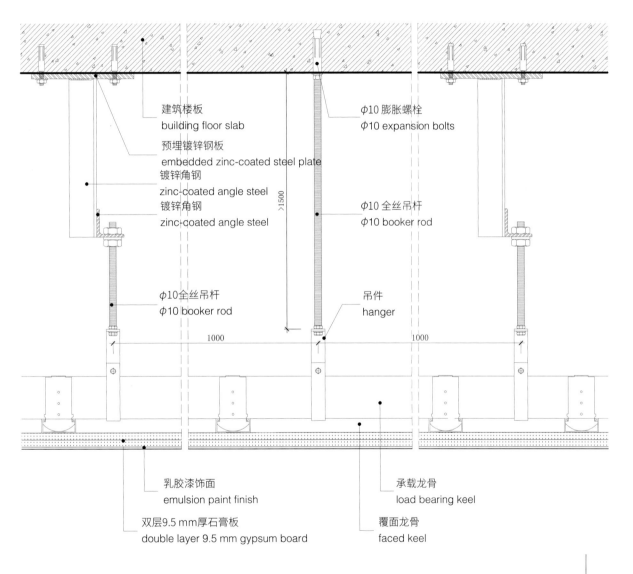

节点图　DETAIL

挡烟垂壁天花
Smoke Restraining Screen Ceiling

132P / 133P

重点 / KEY POINTS

挡烟垂壁主要用于高层或超高层大型商场、写字楼以及仓库等场所，能有效阻挡烟雾在建筑顶棚下的横向流动，以利提高在防烟分区内的排烟效果。自楼板下垂500 mm，材质一般为玻璃。

Smoke restraining screen is mainly applied in buildings including high-rise and super high-rise shopping malls, office buildings, warehouses, and so on. It can effectively stop smoke from lateral flowing under ceilings so as to promote smoke extraction in smoke bays. It can hang down from floor slab for 500 mm. It is mostly made of glasses.

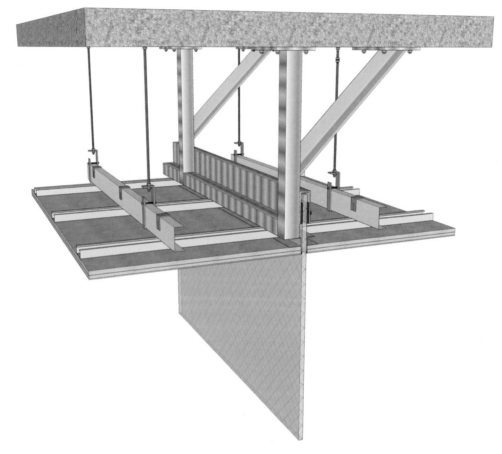

微信扫码，了解更多关于"防火规范"的知识

微信扫码，观看"挡烟垂壁天花"三维节点动图

吊顶工艺节点　DETAILS OF SUSPENDED CEILING PROCESSING TECHNIQUES

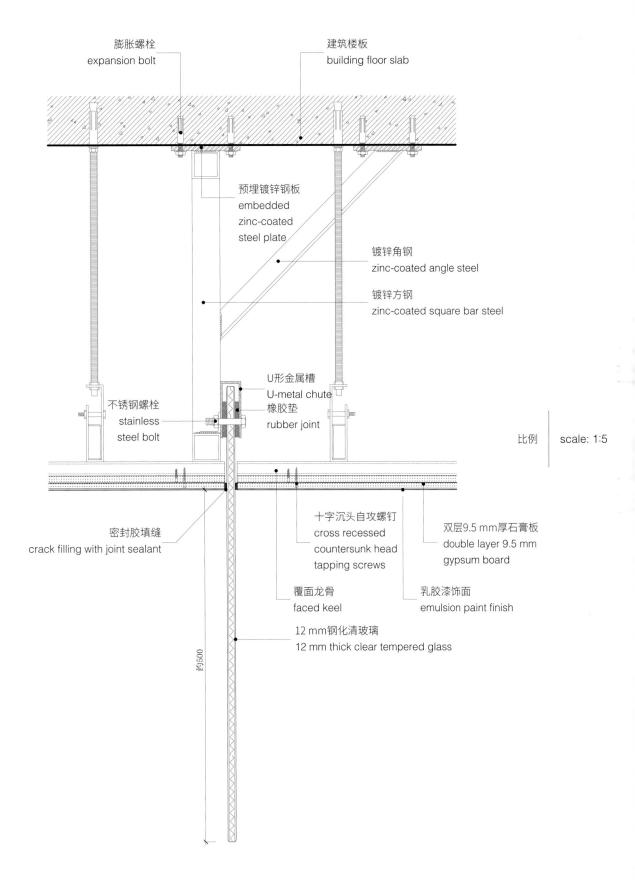

节点图　DETAIL

可升降挡烟垂壁天花
Lifting Smoke Restraining Screen Ceiling

134P / 135P

重点 / KEY POINTS

挡烟垂壁与烟感探测联动，当烟感器报警后，挡烟垂壁自动下降至挡烟工作位置，材质一般为无机布。

Smoke restraining screen ceiling are linked with smoke detectors. As the latter set off the alarm, smoke restraining screen automatically get down to the smoke stop position. The smoke stop rolling screen is generally made from inorganic fabric.

微信扫码，观看"可升降挡烟垂壁天花"三维节点动图

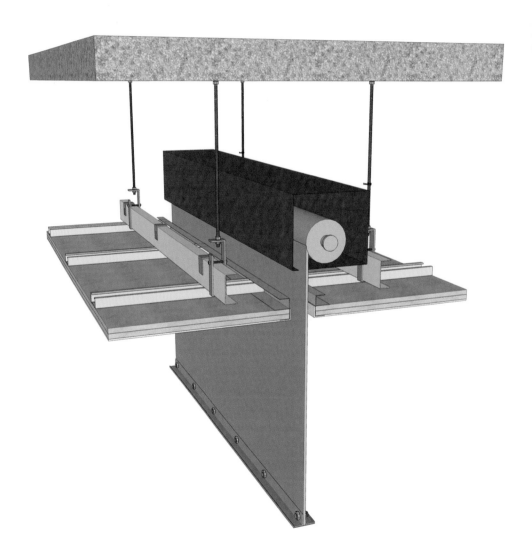

三维图　PERSPECTIVE

吊顶工艺节点　DETAILS OF SUSPENDED CEILING PROCESSING TECHNIQUES

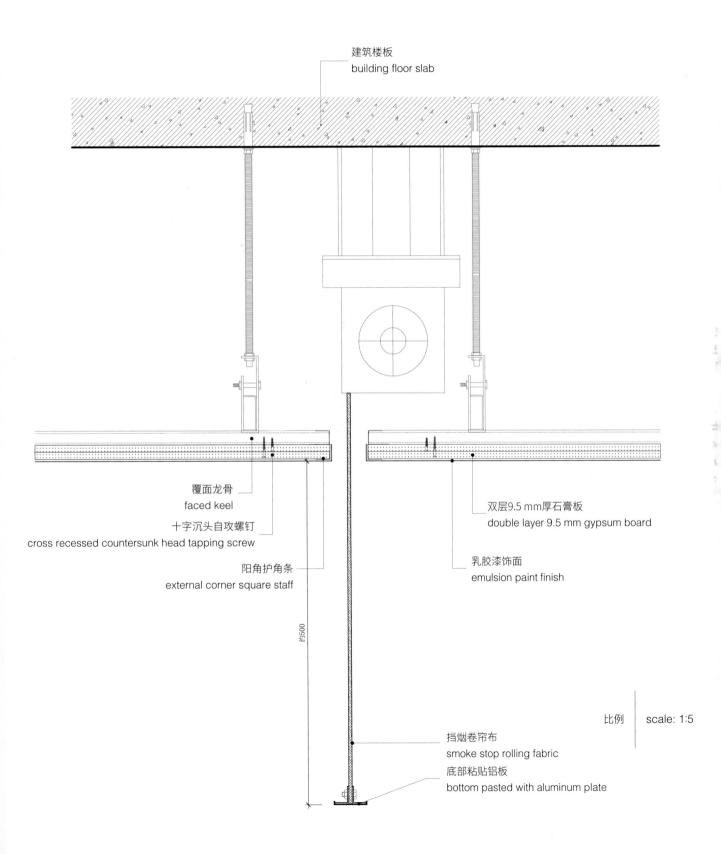

节点图　DETAIL

单轨钢制防火卷帘
Monorail Steel Fire Shutter

136P / 137P

重点 / KEY POINTS

单轨钢制卷帘是最常见的卷帘形式，安装简单，和装饰天花的处理也先对容易。单轨钢制卷帘无法做成弧线形式。

As the most common rolling fabric, monorail steel fire shutter is easy to be equipped. Construction between monorail steel fire shutter and decorative railing is relatively easy. It cannot be made into arc.

微信扫码，了解更多关于"防火规范"的知识

微信扫码，观看"单轨钢制防火卷帘"三维节点动图

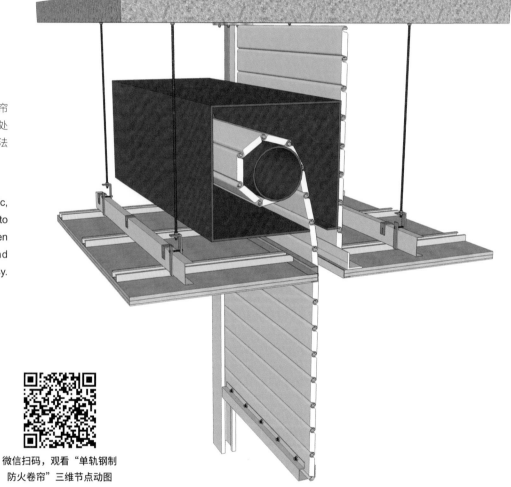

三维图　PERSPECTIVE

吊顶工艺节点　DETAILS OF SUSPENDED CEILING PROCESSING TECHNIQUES

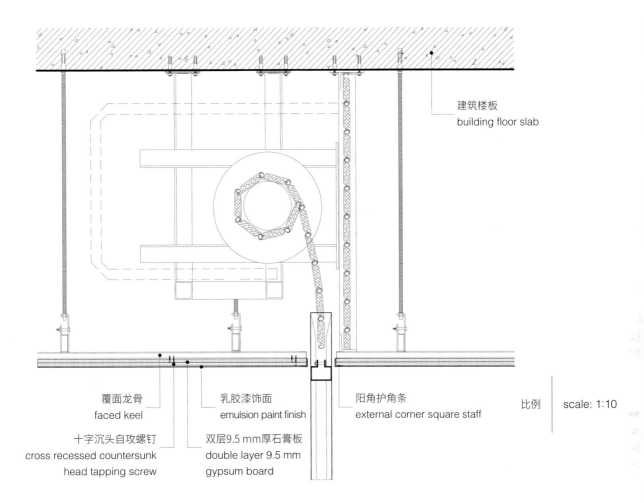

比例　scale: 1:10

- 建筑楼板 building floor slab
- 覆面龙骨 faced keel
- 乳胶漆饰面 emulsion paint finish
- 阳角护角条 external corner square staff
- 十字沉头自攻螺钉 cross recessed countersunk head tapping screw
- 双层9.5 mm厚石膏板 double layer 9.5 mm gypsum board

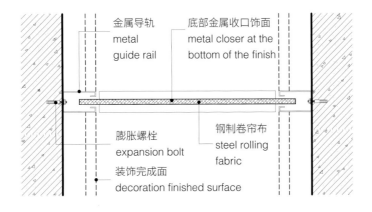

- 金属导轨 metal guide rail
- 底部金属收口饰面 metal closer at the bottom of the finish
- 膨胀螺栓 expansion bolt
- 钢制卷帘布 steel rolling fabric
- 装饰完成面 decoration finished surface

节点图　DETAIL

双轨无机布防火卷帘
Double-Track Inorganic Rolling Fabric

138P / 139P

重点 / KEY POINTS

单轨钢制防火卷帘适用于直线安装，双轨无机布防火卷帘适用于对抗风压要求低的场所。卷帘自重轻，可减少建筑载荷。特级无机复合防火卷帘门面1.2 mm厚双轨、双帘面，其安全性更可靠。

Monorail steel fire shutter is suitable for straight-line equipment, while double-track inorganic rolling fabric is suitable for buildings with lower level of wind pressure. Rolling fabric has lower deadweight, lightening loads on the structure. Super inorganic rolling fabric has a width of 1.2 mm thick double-track and double-curtain, which has reliable security.

微信扫码，观看"双轨无机布防火卷帘"三维节点动图

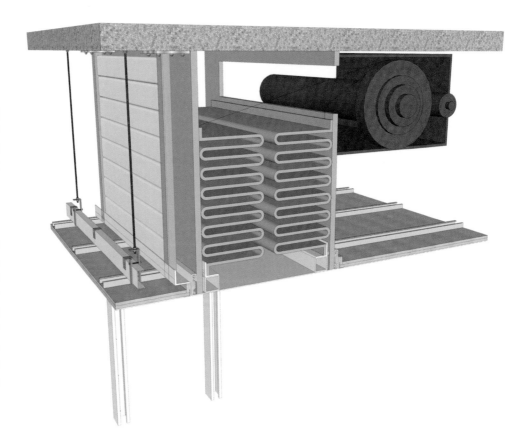

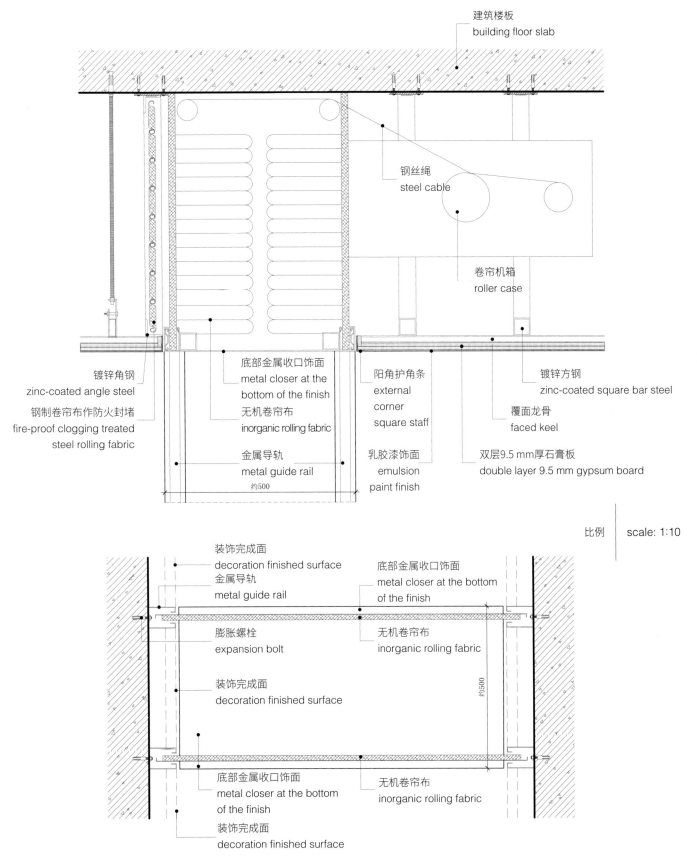

3

地坪工艺节点
DETAILS OF FLOOR PROCESSING TECHNIQUES

本篇的主要内容是不同地面材料的工艺做法，涉及的材料有石材、水磨石、木地板、地坪漆、地毯等，还包括防水处理及地坪抬高的常规做法。

不管何种地坪材质，其工艺构造基本都可以概括为结构层、找平层、结合层、饰面层四部分。希望大家在设计过程和施工现场中可以发现其中的规律并合理利用。

石材/瓷砖地坪　|干铺法|
Stone/Ceramic Tile Floor　| Dry Laid Method |

142P ↗　143P ↗

重点 / KEY POINTS

　　石材干铺法：采用 1:3 干硬性水泥砂浆打底，再在石材背面刮满专用黏结剂，然后把石材铺装在干硬性水泥砂浆上。

Stone dry laid method: render with the 1:3 harsh cement mortar, and then pave specified adhesive on the stone, finally lay the stone on the harsh cement mortar.

重点 / KEY POINTS

　　水磨石是采用透明玻璃、陶瓷颗粒、金属颗粒、贝壳、石英石等骨料与高分子树脂相混合，并经现场浇筑（摊铺）、研磨、抛光等工艺而打造出的一种整体无缝同质同心装饰地材。

Terrazzo is a seamless homogenous and concentric decorative flooring that is comprised of polymer resins and many kinds of aggregate such as clear glass, ceramic particles, metal particles, shell, quartz stone, etc. Processes of on-site pouring(paving), grinding, polishing are also used.

水磨石地坪
Terrazzo Floor

142P ↘　143P ↘

微信扫码，了解更多关于"石材地坪"的知识

微信扫码，观看"石材/瓷砖地坪|干铺法|"三维节点动图

微信扫码，观看"水磨石地坪"三维节点动图

微信扫码，了解更多关于"水磨石"的知识

地坪工艺节点　DETAILS OF FLOOR PROCESSING TECHNIQUES

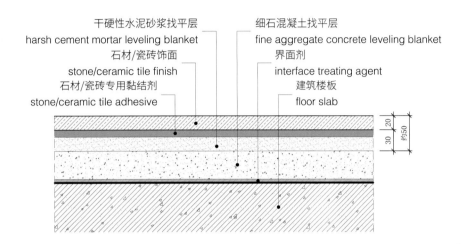

比例　scale: 1:5

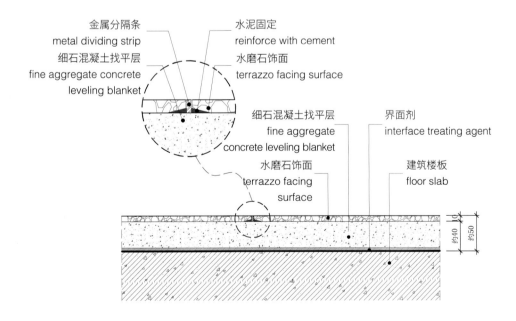

节点图　DETAIL

木地板地坪　｜混凝土基层｜
Wood Floor　｜Concrete Base｜

144P ↗　145P ↗

重点 / KEY POINTS

选择复合地板或实木复合地板时可以采用这种铺装方式。

This paving method is suitable for compound floor or parquet.

微信扫码，了解更多关于"木地板"的知识

微信扫码，观看"木地板地坪｜混凝土基层｜"三维节点动图

微信扫码，观看"木地板地坪｜木龙骨基层｜"三维节点动图

重点 / KEY POINTS

选择实木地板或高级实木复合地板时可以采用这种铺装方式。

This paving method is suitable for wood floor or high quality wood compound floor.

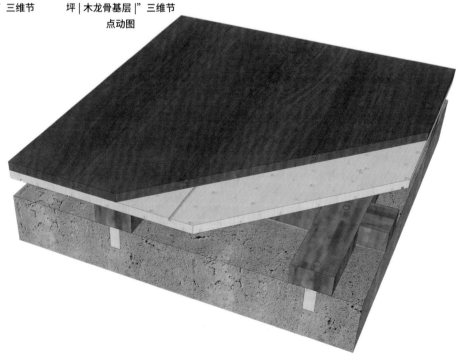

木地板地坪　｜木龙骨基层｜
Wood Floor　｜Wooden Joist Base｜

144P ↘　145P ↘

地坪工艺节点　DETAILS OF FLOOR PROCESSING TECHNIQUES

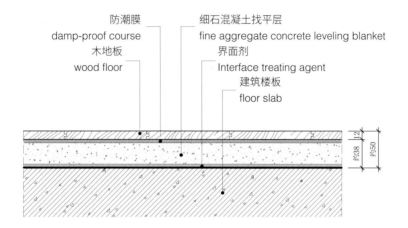

防潮膜　damp-proof course
木地板　wood floor
细石混凝土找平层　fine aggregate concrete leveling blanket
界面剂　Interface treating agent
建筑楼板　floor slab

比例　scale: 1:5

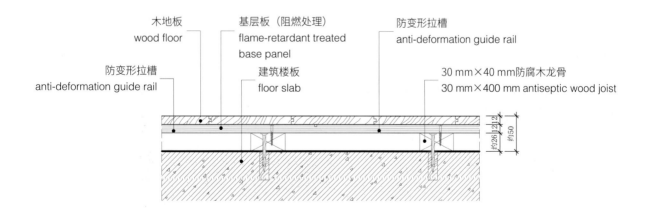

木地板　wood floor
基层板（阻燃处理）　flame-retardant treated base panel
防变形拉槽　anti-deformation guide rail
防变形拉槽　anti-deformation guide rail
建筑楼板　floor slab
30 mm×40 mm 防腐木龙骨　30 mm×400 mm antiseptic wood joist

节点图　DETAIL

145

防腐木地坪
Antiseptic Wood Floor

146P ↗　147P ↗

重点 / KEY POINTS

　　防腐木是将普通木材经过化学处理后得到的，其稳定性高，多用于室外。

Antiseptic wood is chemically treated plain wood. It has high stability and is often used outdoors.

微信扫码，了解更多关于"户外木地板"的知识

微信扫码，观看"防腐木地坪"三维节点动图

微信扫码，观看"环氧地坪"三维节点动图

重点 / KEY POINTS

　　环氧地坪漆主要成分是环氧树脂和固化剂，多用于车库及创意型室内空间。

The main compositions of epoxy floor paint, often used in garage or creative interior space, are epoxy resin and curing agent.

环氧地坪
Epoxy Floor

146P ↘　147P ↘

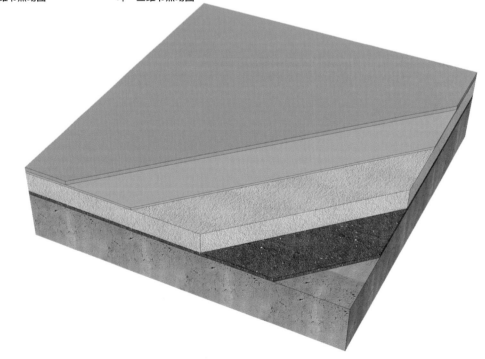

地坪工艺节点　DETAILS OF FLOOR PROCESSING TECHNIQUES

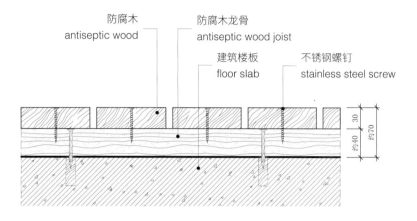

比例 | scale: 1:5

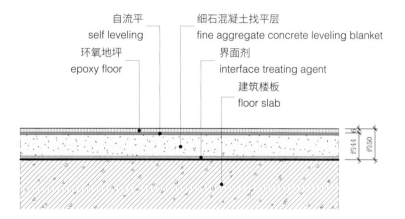

节点图　DETAIL

架空地板地坪
Aerial Floor

148P ↗   149P ↗

重点 / KEY POINTS

架空地板为成品，高度不宜小于 100 mm，用于地面穿管布线，铺设快捷，便于检修，多用于大空间办公。

Finished aerial floor is used for pipelines and wires, it can be quickly deployed and easily overhauled and often used in open bullpen office. In addition, the height of aerial floor should not be less than 100 mm.

微信扫码，了解更多关于"架空地板"的知识

重点 / KEY POINTS

玻璃架空地坪往往和灯光效果结合，要注意灯具的安装及检修问题。

Glass aerial floor is usually combined with light effect. The Installation and maintenance of luminaire should be paid attention.

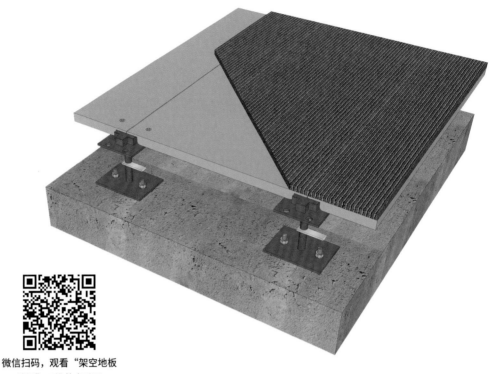

微信扫码，观看"架空地板地坪"三维节点动图

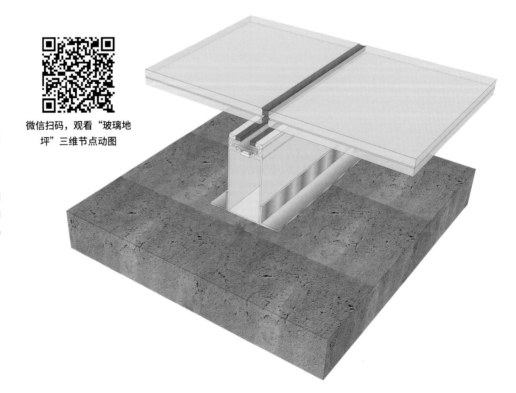

微信扫码，观看"玻璃地坪"三维节点动图

玻璃地坪
Glass Floor

148P ↘   149P ↘

地坪工艺节点　DETAILS OF FLOOR PROCESSING TECHNIQUES

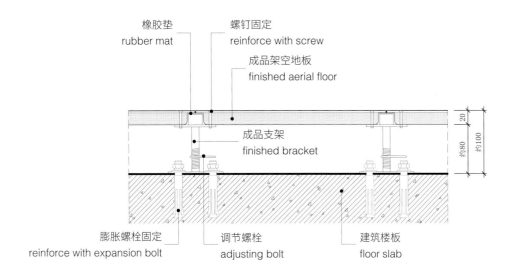

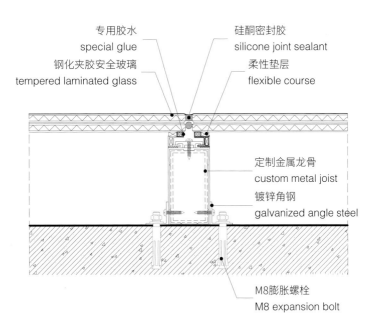

比例　scale: 1:5

节点图　DETAIL

149

块毯地坪
Area Rugs

150P ↗ 151P ↗

重点 / KEY POINTS

块毯常用规格为 500 mm×500 mm，多用于办公场所，直接粘贴，不需要倒刺条。

The common specification of area rugs is 500 mm × 500 mm. It is often used workplace and can be paste directly without tack strip.

微信扫码，了解更多关于"方块地毯"的知识

微信扫码，观看"块毯地坪"三维节点动图

微信扫码，观看"满铺地毯地坪"三维节点动图

重点 / KEY POINTS

满铺地毯多用于宴会厅、贵宾厅等空间，安装时在边缘处需要设置倒刺条固定.

Wall-to-wall carpeting is often used in banquet hall and VIP lounge. Tack strip to be set on the edge is necessary while installation.

满铺地毯地坪
Wall-to-Wall Carpeting Floor

150P ↘ 151P ↘

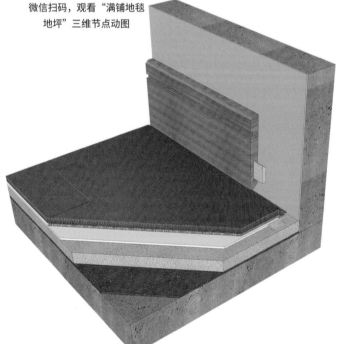

地坪工艺节点　DETAILS OF FLOOR PROCESSING TECHNIQUES

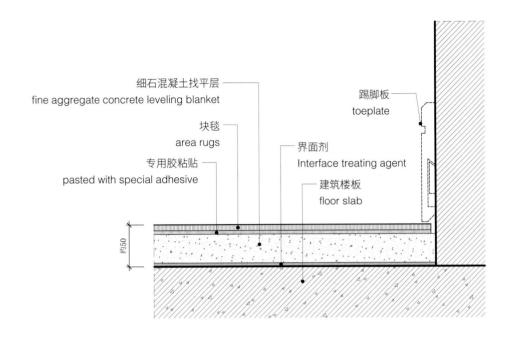

比例　scale: 1:5

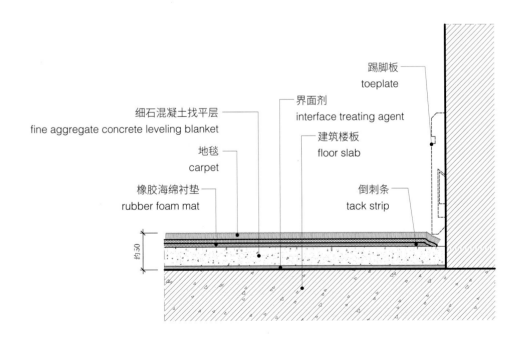

节点图　DETAIL

塑胶地板地坪
Plastic Floor

152P ↗   153P ↗

重点 / KEY POINTS

塑胶地板（PVC 地板）是一种人造材料，有卷材和块材两种形式，多应用于教育、办公、医疗等项目。

Plastic floor(PVC floor) is a kind of artificial material with the shape in coil and bulk. It is often used in education, office and medical projects.

微信扫码，了解更多关于"塑胶地板"的知识

微信扫码，观看"塑胶地板地坪"三维节点动图

微信扫码，观看"地暖地坪"三维节点动图

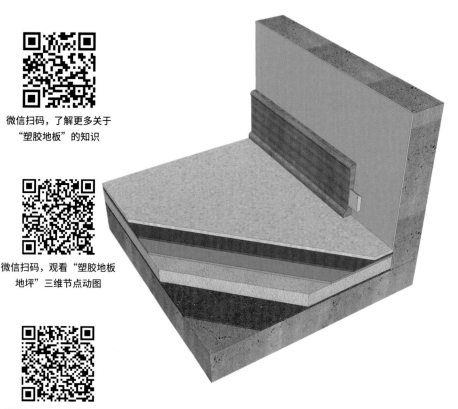

重点 / KEY POINTS

图中的地暖是指水暖形式，地暖区域的地面材料通常选用石材、地砖或木地板。在选择木地板时一定要注意是否为地暖木地板，否则地暖会导致普通木地板起拱变形。

The floor heating in the picture is water heating. Stone, tile and wood are common materials in the floor heating area. Wood floor special for floor heating rather than plain floor must be chosen to avoid bagging deformation due to the heat.

地暖地坪
Floor Heating Floor

152P ↘   153P ↘

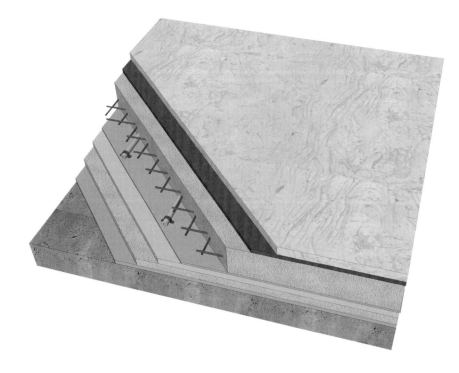

地坪工艺节点　DETAILS OF FLOOR PROCESSING TECHNIQUES

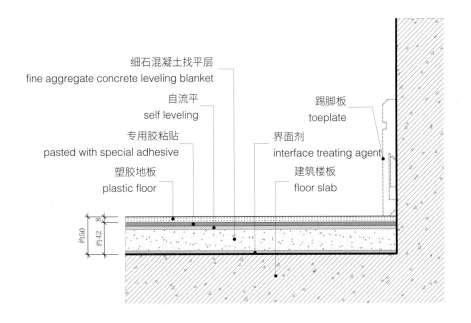

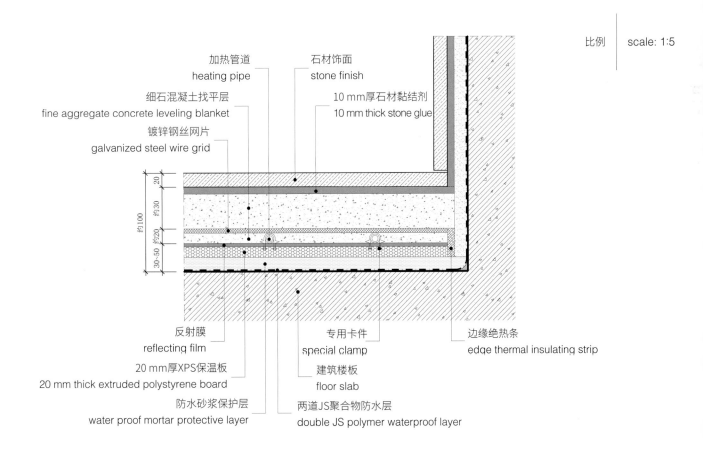

比例　scale: 1:5

节点图　DETAIL

砌筑地台
Masonry Platform

154P ↗   155P ↗

重点 / KEY POINTS

砌筑地台应用于地台高度较低（300~500 mm）或地面局部抬高时。

Masonry platform is used when the platform is low(about 300~500 mm) or ground is elevated partially.

微信扫码，观看"砌筑地台"三维节点动图

微信扫码，观看"钢架地台"三维节点动图

重点 / KEY POINTS

地台高度较高，多用于阶梯教室、报告厅等，钢结构的尺寸及搭接方式需要根据实际情况具体分析。

This kind of masonry platform is used when the platform is high, such as lecture theatre and lecture hall, etc. The size and building method of steel structure is different on each specific case.

钢架地台
Steel Structure Platform

154P ↘   155P ↘

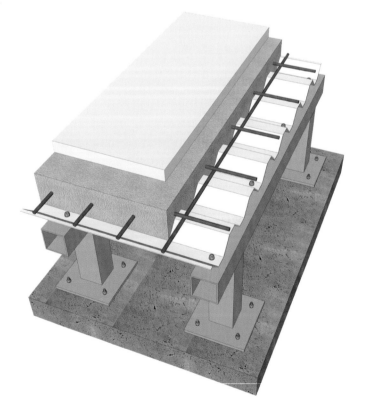

三维图　PERSPECTIVE

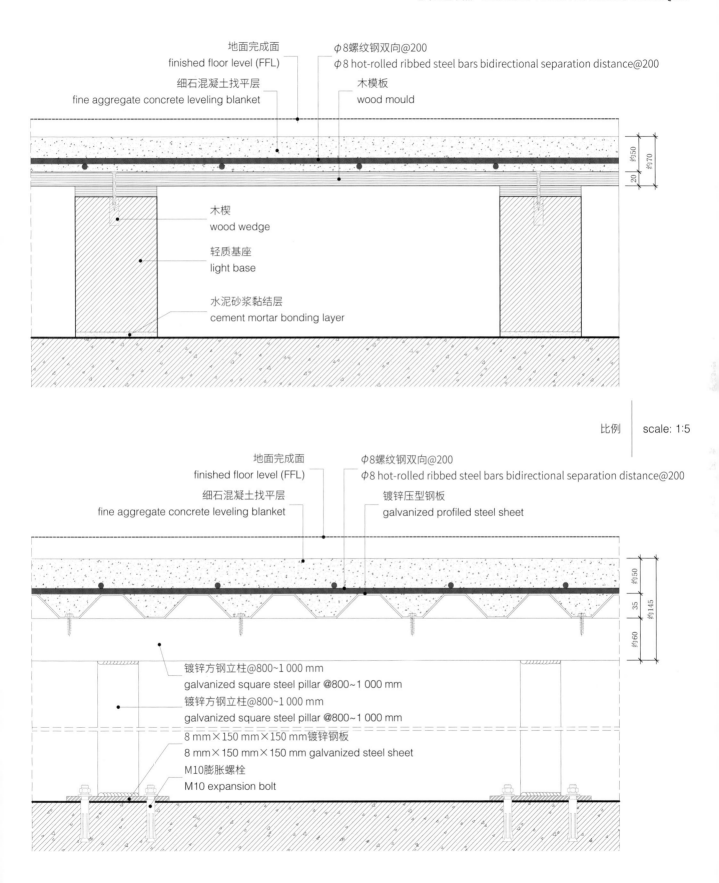

石材 - 木地板交接地坪
Stone-Wood Floor Transition

156P ↗   157P ↗

重点 / KEY POINTS

不同地面材质之间使用金属嵌条过渡是比较常见的做法，也可以不做嵌条，但是两种不同的材料平接对工艺及收口要求会更高。

Medal strip is often used for different floor materials' transition. Without strip, it would be more demanding for processing and binding to make one terminate flush with the other.

微信扫码，观看"石材 - 木地板交接地坪"三维节点动图

微信扫码，观看"石材 - 满铺地毯交接地坪"三维节点动图

重点 / KEY POINTS

金属嵌条可以固定在地面，也可以粘贴在石材侧面。

Medal strip could be fixed on the ground, or be pasted on the side of the stone.

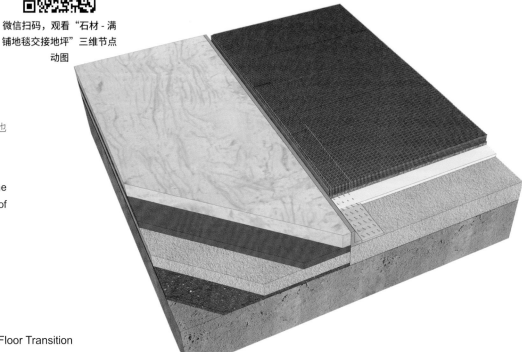

石材 - 满铺地毯交接地坪
Stone - Wall-to-Wall Carpeting Floor Transition

156P ↘   157P ↘

地坪工艺节点　DETAILS OF FLOOR PROCESSING TECHNIQUES

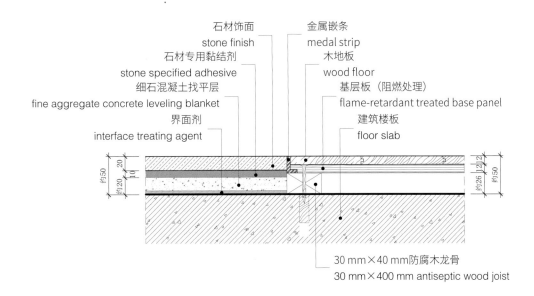

scale: 1:5

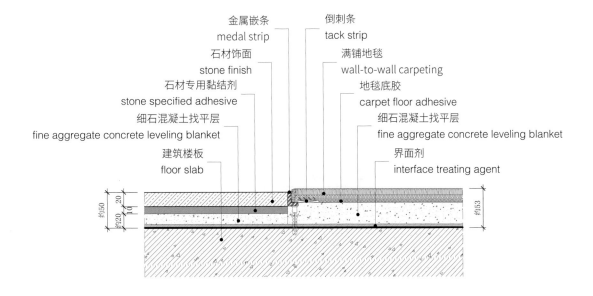

节点图　DETAIL

石材 - 除泥垫交接地坪
Stone - Silt Removal Mat Floor Transition

158P ↗ 159P ↗

重点 / KEY POINTS

除泥垫是成品，多用于公共空间入口处，可有效地刮除泥尘和水分，保持室内整洁。

Silt removal mat is finish product, often used in the entrance of public area. It can scrape silt and water efficiently and keep the room clean.

微信扫码，观看"石材 - 除泥垫交接地坪"三维节点动图

微信扫码，观看"木地板 - 满铺地毯交接地坪"三维节点动图

重点 / KEY POINTS

不同地面材质之间使用金属嵌条过渡是比较常见的做法。

Medal strip is often used for different floor materials' transition.

木地板 - 满铺地毯交接地坪
Stone - Wall-To-Wall Carpeting Floor Transition

158P ↘ 159P ↘

地坪工艺节点　DETAILS OF FLOOR PROCESSING TECHNIQUES

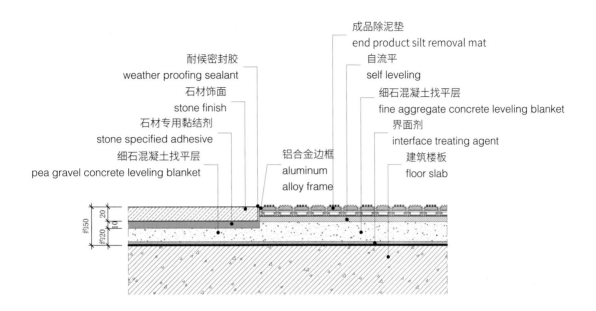

比例　scale: 1:5

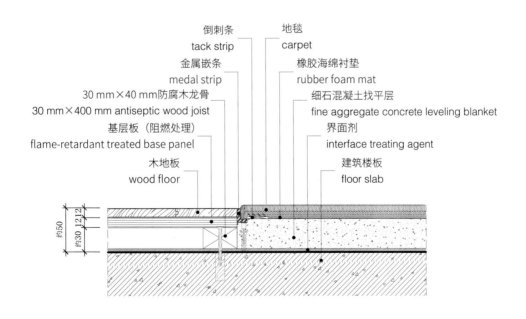

节点图　DETAIL

地坪 - 幕墙收口　|高|
Floor-Curtain Wall Closure |High|

160P / 161P

重点 / KEY POINTS

当幕墙横档结构高于楼面地坪完成面时，需要通过设置窗台造型和幕墙结构相结合，窗台内部构造不可和幕墙结构有硬性连接。

When the curtain wall transverse structure is higher than the floor finish surface, the windowsill shape should be set combined with the curtain wall structure. The interior structure of the windowsill cannot be rigidly connected with the curtain wall structure.

微信扫码，观看
"地坪 - 幕墙收口 | 高 |"
三维节点动图

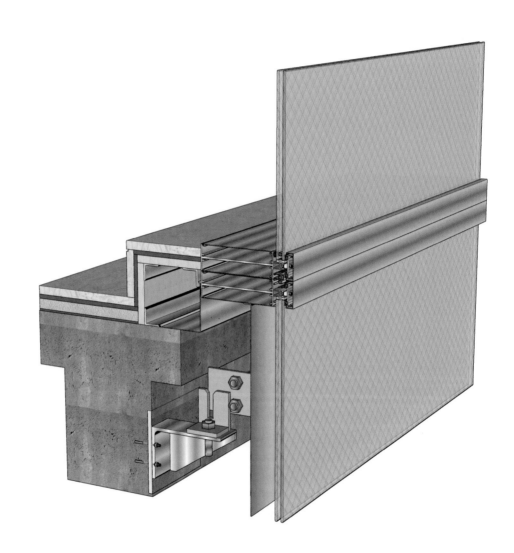

三维图　PERSPECTIVE

地坪工艺节点　DETAILS OF FLOOR PROCESSING TECHNIQUES

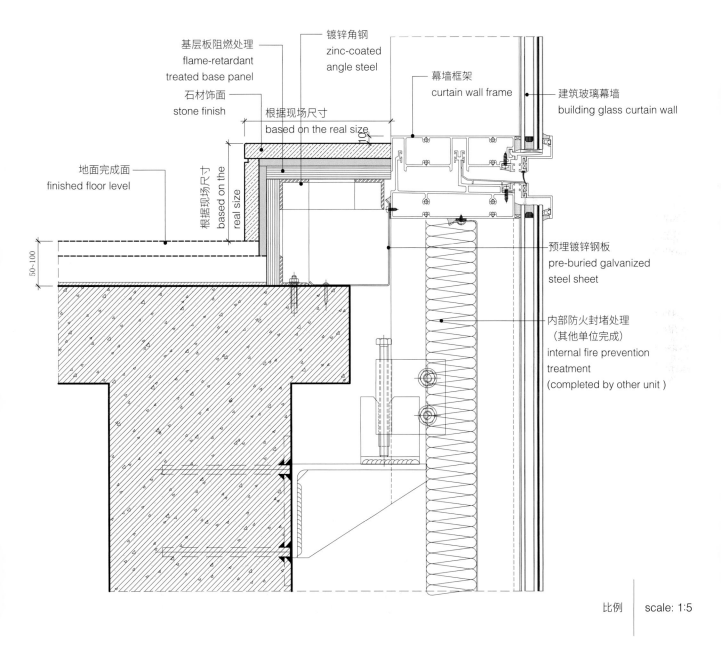

节点图　DETAIL

地坪 - 幕墙收口　|平|
Floor-Curtain Wall Closure　|Flat|

162P / 163P

重点 / KEY POINTS

当幕墙横档结构和楼面地坪完成面基本持平时，需要在地坪和幕墙结构连接处进行打胶密封处理。

When the curtain wall transverse structure and the floor finish surface are basically the same, the joint of floor and curtain wall structure needs to be sealed.

微信扫码，观看
"地坪 - 幕墙收口 | 平 |"
三维节点动图

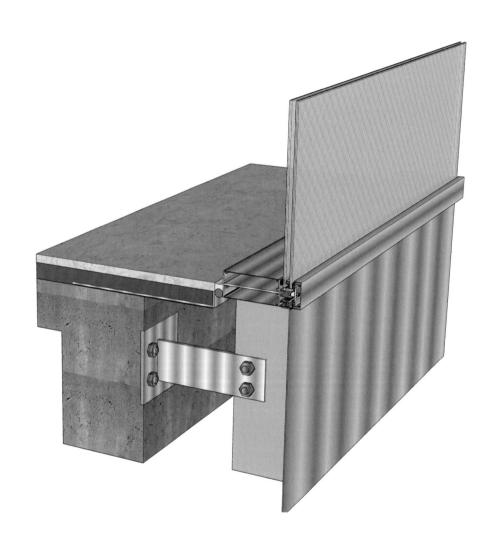

三维图　PERSPECTIVE

地坪工艺节点　DETAILS OF FLOOR PROCESSING TECHNIQUES

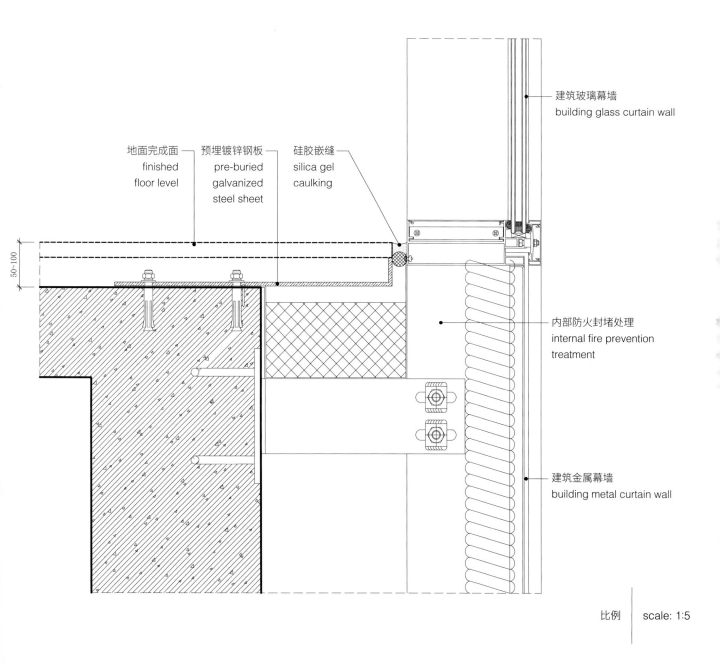

节点图　DETAIL

卫生间淋浴房挡水槛地坪（铺法一）
Bathroom or Shower Room Water Retaining Sill Floor (Laying Method 1)

164P ↗   165P ↗

重点 / KEY POINTS

淋浴房挡水处需做导墙。

Diverting wall is necessary in the water retaining area of shower room.

微信扫码，观看"卫生间淋浴房挡水槛地坪（铺法一）"三维节点动图

微信扫码，了解更多关于"淋浴间"的知识

微信扫码，观看"卫生间淋浴房挡水槛地坪（铺法二）"三维节点动图

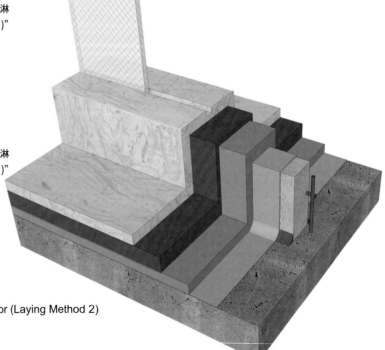

卫生间淋浴房挡水槛地坪（铺法二）
Bathroom or Shower Room Water Retaining Sill Floor (Laying Method 2)

164P ↘   165P ↘

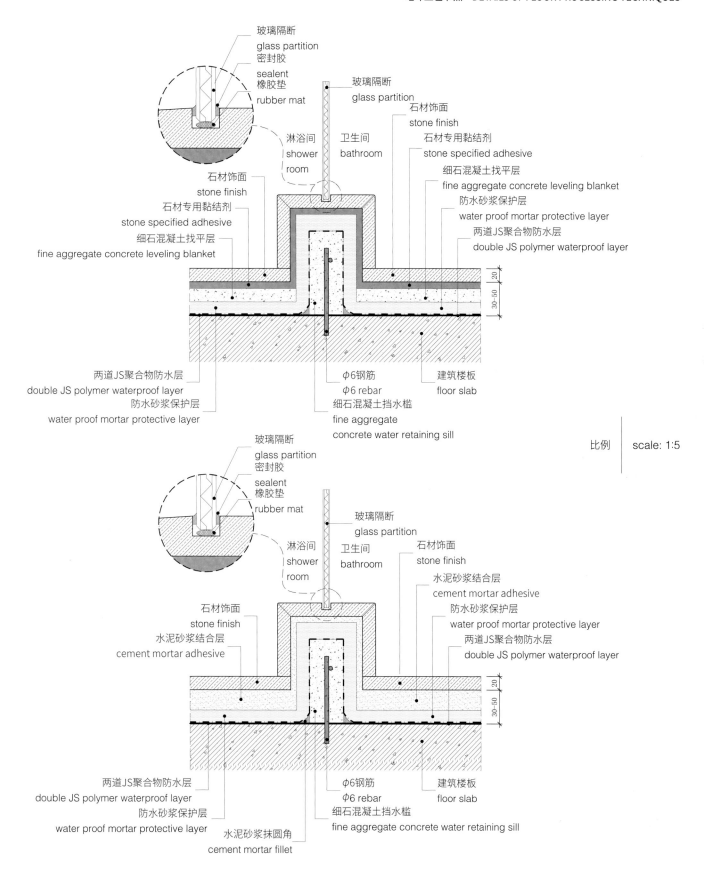

卫生间地漏　|明装式|
Toilet Floor Drain　|Opening|

166P ↗　　167P ↗

重点 / KEY POINTS

常规地漏安装方式便于施工和检修。

The conventional floor drain installation method is convenient for construction and maintenance.

微信扫码，了解更多关于"卫生间地漏"的知识

微信扫码，观看"卫生间地漏 | 明装式 |"三维节点动图

微信扫码，观看"卫生间地漏 | 隐藏式 |"三维节点动图

重点 / KEY POINTS

地漏设置于活动检修盖板下，可以使地坪装饰效果保持完整，同时满足功能需求。需要设置排水沟集中排水，对地坪完成面高度要求较大。

The floor drain is arranged under the mobile maintenance cover plate to ensure the integrity of the floor and meet the functional requirements. Gutters need to be set up for centralized drainage, and there is a great need of floor completion height.

卫生间地漏　|隐藏式|
Toilet Floor Drain　|Hidden|

166P ↘　　167P ↘

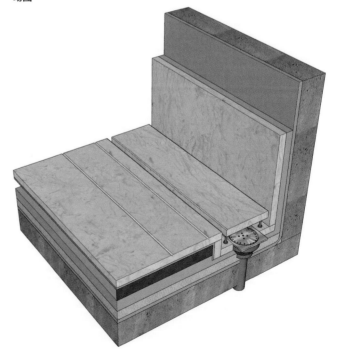

三维图　PERSPECTIVE

地坪工艺节点　DETAILS OF FLOOR PROCESSING TECHNIQUES

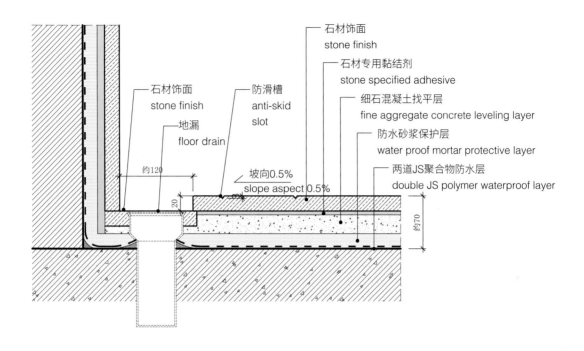

scale: 1:5

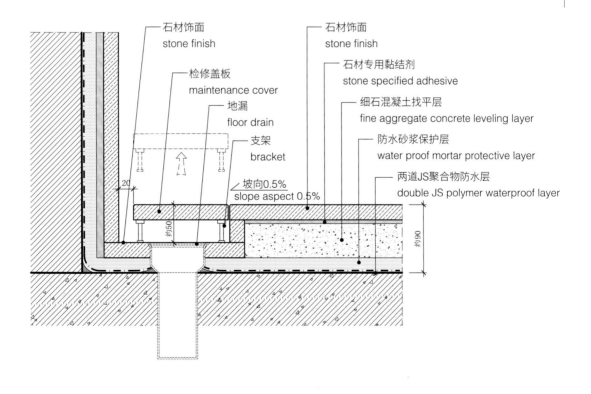

节点图　DETAIL

卫生间门槛石地坪（铺法一）
Bathroom Stone Door Sill Floor (Laying Method 1)

168P ↗ 169P ↗

重点 / KEY POINTS

卫生间区域一般会有 20 mm 的降低，但在门槛处最好加做挡水槛，用以分隔干湿区。

The bathroom area is generally lower than elevation by 20 mm. While it would be better to have a water retaining step to separate arid area and humid area.

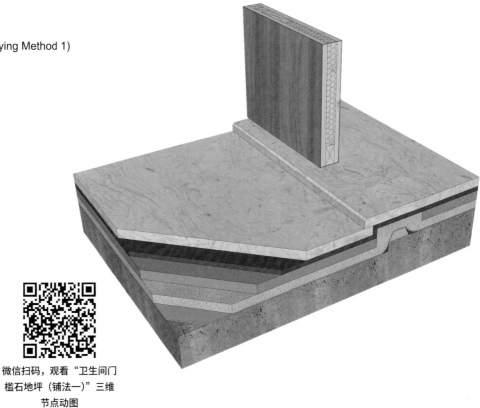

微信扫码，观看"卫生间门槛石地坪（铺法一）"三维节点动图

微信扫码，观看"卫生间门槛石地坪（铺法二）"三维节点动图

卫生间门槛石地坪（铺法二）
Bathroom Stone Door Sill Floor (Laying Method 2)

168P ↘ 169P ↘

地坪工艺节点　DETAILS OF FLOOR PROCESSING TECHNIQUES

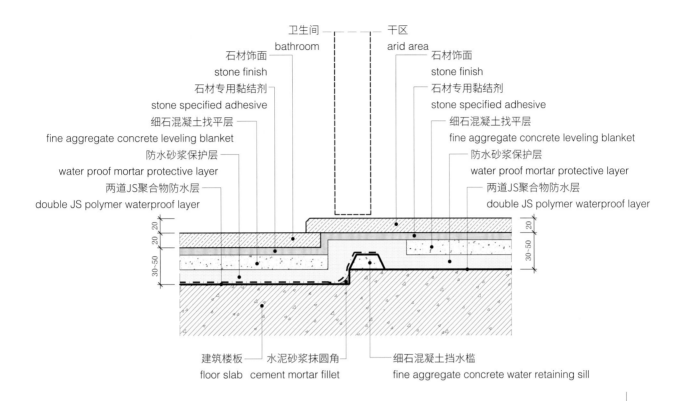

节点图　DETAIL

卫生间玻璃隔断墙面收口
Bathroom Glazed Partition Wall Binding

170P ↗   171P ↗

重点 / KEY POINTS

玻璃隔断的常见厚度为 8 mm、10 mm，必须钢化处理。

The thickness of glazed partition is usually 8 mm or 10 mm. Tempering treatment is necessary.

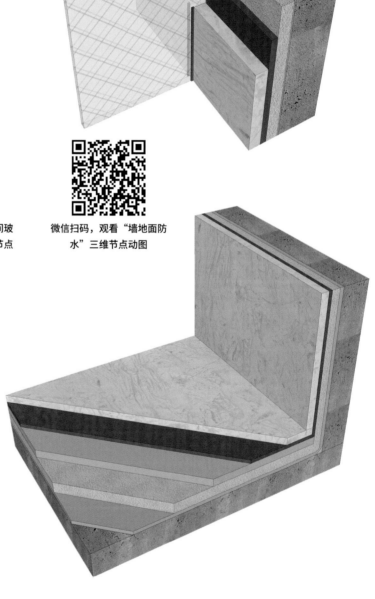

微信扫码，了解更多关于"室内防水"的知识

微信扫码，观看"卫生间玻璃隔断墙面收口"三维节点动图

微信扫码，观看"墙地面防水"三维节点动图

重点 / KEY POINTS

卫生间湿区墙面防水高度为 1 800 mm。

The height for water retaining of wall of bathroom humid area is 1 800 mm.

墙地面防水
Water Proofness of Wall and Floor

170P ↘   171P ↘

地坪工艺节点　DETAILS OF FLOOR PROCESSING TECHNIQUES

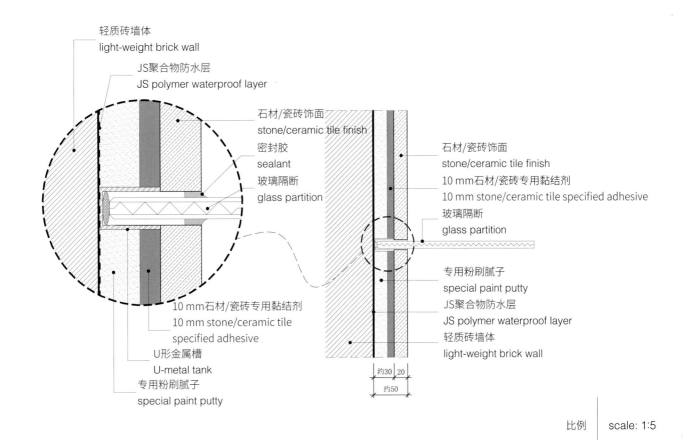

比例　scale: 1:5

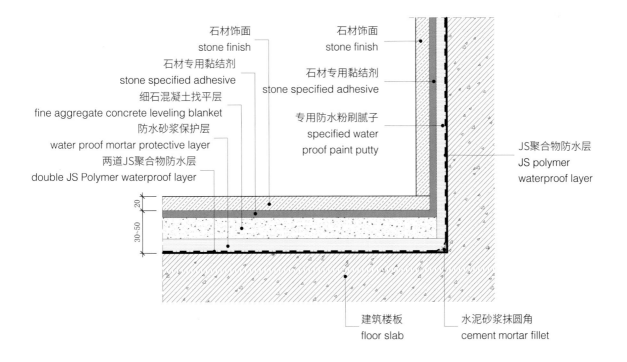

节点图　DETAIL

地面变形缝
Ground Deformation Joint

172P ↗   173P ↗

重点 / KEY POINTS

由于地基不均匀沉降、温度变化、地震等因素的影响，建筑容易发生变形或破坏，因此在建筑设计时将房屋划分成若干个独立部分，使各部分能独立自由地变化，这种将建筑物垂直分开的预留缝称为变形缝。变形缝盖板是成熟产品，用于变形缝的遮蔽和保护。

图中所示的地面变形缝盖板可在中心板区域嵌装和地坪相同的饰面材料。墙面变形缝处理则是通过现场施工手法解决。

贯穿变形缝两侧的构件不能硬性连接，需要满足足够的位移要求。

There are many factors that can make the building deform or destroyed, which including the uneven settlement of foundation, temperature change, earthquakes and so on. Thus, the buildings need to be divided into several independent parts in the architectural design, which can make each part change independently and freely. This reserved joint that separates the building vertically is called deformation joint. The deformation joint cover plate is a mature product, which is used to shield and protect the deformation joint.

The ground deformation joint cover plate shown in the figure can be embedded in the central plate area with the same decorative material as the floor, while the treatment of wall deformation joint is solved by means of site-construction.

The components on both sides of the deformation joint cannot be rigidly connected, and they need enough displacement.

墙面变形缝
Wall Deformation Joint

172P ↘   173P ↘

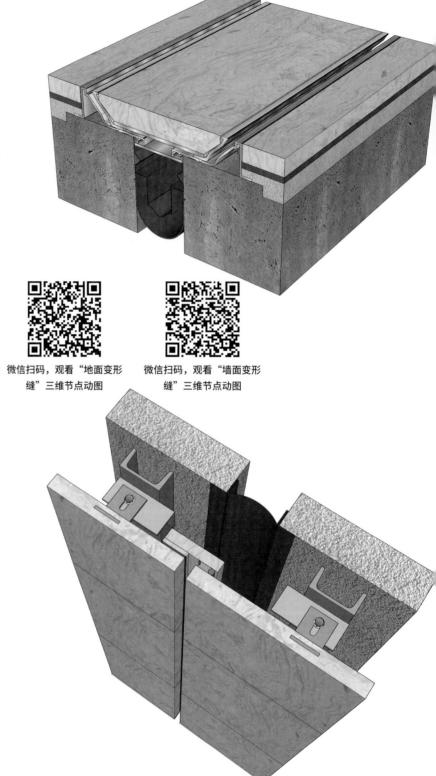

微信扫码，观看"地面变形缝"三维节点动图

微信扫码，观看"墙面变形缝"三维节点动图

地坪工艺节点　DETAILS OF FLOOR PROCESSING TECHNIQUES

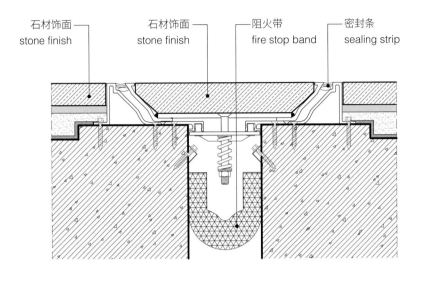

比例 | scale: 1:5

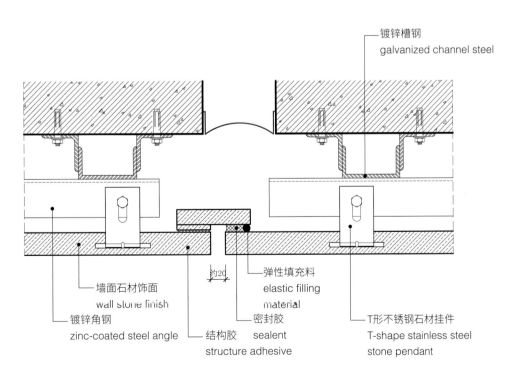

节点图　DETAIL

泳池隐藏式排水沟
Hidden Drainage Ditch in Swimming Pool

174P / 175P

重点 / KEY POINTS

这是一种对排水沟的常见装饰处理手法，在地坪上看不到排水格栅，保证了地坪效果的完整，通过活动盖板来进行维护检修。

It is a common decorative treatment method for drainage ditches. No drainage grille can be seen on the floor, which ensures the integrity of the floor. Moreover, the drainage ditch can be maintained and repaired through movable cover plate.

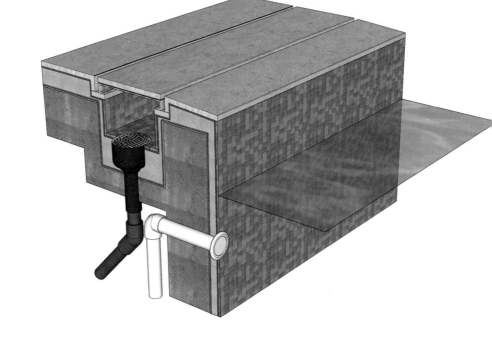

微信扫码，观看"泳池隐藏式排水沟"三维节点动图

三维图　PERSPECTIVE

地坪工艺节点　DETAILS OF FLOOR PROCESSING TECHNIQUES

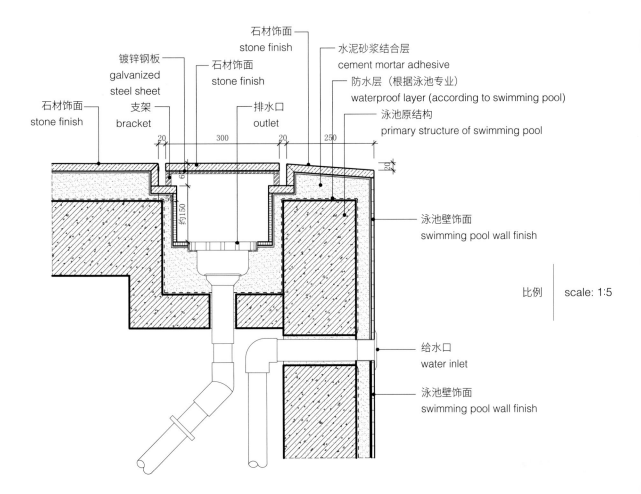

节点图　DETAIL

泳池明装式排水沟
Open Drainage Ditch in Swimming Pool

176P / 177P

重点 / KEY POINTS

这是排水沟常规做法。明装金属格栅盖板具有排水顺畅、施工便捷的优点。

This is a routine method for drainage ditch, with the open metal grille cover, smooth drainage, and convenient construction.

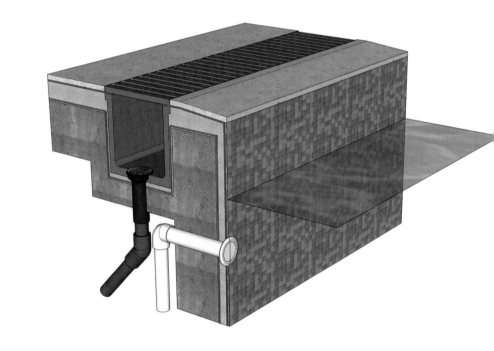

微信扫码，观看"泳池明装式排水沟"三维节点动图

地坪工艺节点　DETAILS OF FLOOR PROCESSING TECHNIQUES

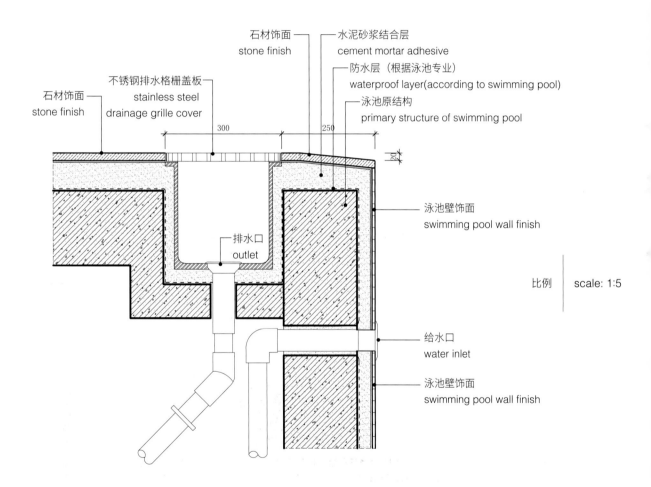

节点图　DETAIL

石材踢脚　|凸|
Stone Baseboard　| Convex |

178P ↗　179P ↗

微信扫码，了解更多关于
"踢脚线"的知识

微信扫码，观看"石材踢脚
|凸|"三维节点动图

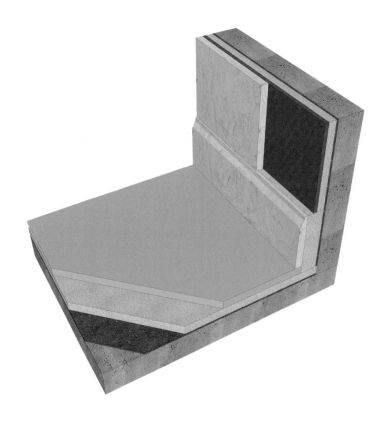

微信扫码，观看"石材踢脚
|平|"三维节点动图

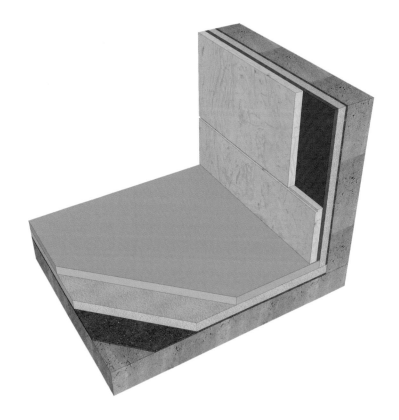

石材踢脚　|平|
Stone Baseboard　| Flat |

178P ↘　179P ↘

地坪工艺节点　DETAILS OF FLOOR PROCESSING TECHNIQUES

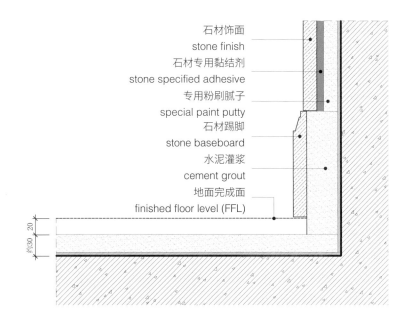

比例　scale: 1:5

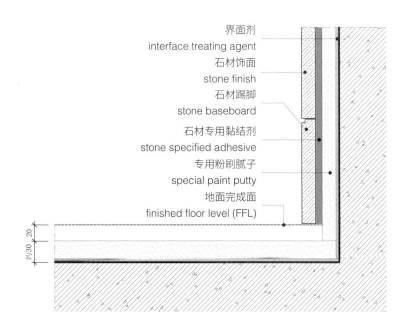

节点图　DETAIL

179

金属踢脚　|凹|
Metal Baseboard　| Concave |

180P ↗　181P ↗

重点 / KEY POINTS

　　内凹式金属踢脚需要一定的墙面完成面才能实现，设计时需要注意。

Here is one thing to note in the design period that concave metal baseboard is based on part of wall finish.

微信扫码，了解更多关于"金属收口"的知识

微信扫码，观看"金属踢脚|凹|"三维节点动图

微信扫码，观看"金属踢脚|凸|"三维节点动图

重点 / KEY POINTS

　　一般采用1 mm 或 1.2 mm 厚不锈钢粘贴在基层板上。

The thickness of stainless steel panels pasted on the baseboard is generally 1 mm or 1.2 mm.

金属踢脚　|凸|
Metal Baseboard　| Convex |

180P ↘　181P ↘

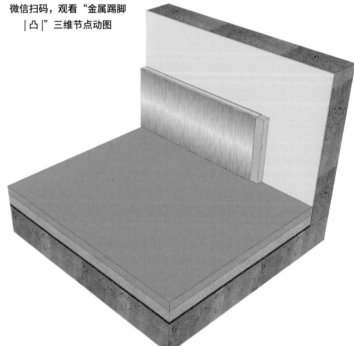

180　　　　　三维图　PERSPECTIVE

地坪工艺节点　DETAILS OF FLOOR PROCESSING TECHNIQUES

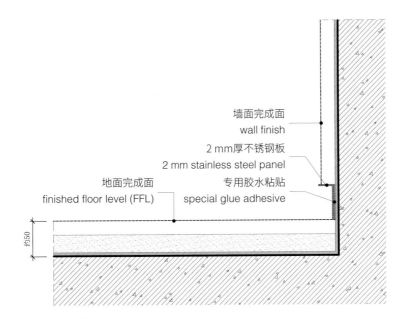

比例　scale: 1:5

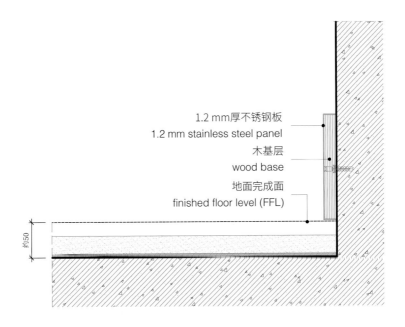

节点图　DETAIL

石材踏步　|混凝土楼梯|
Stone Step　| Concrete Staircase |

182P ↗　183P ↗

重点 / KEY POINTS

石材阳角要做倒角处理，踏步前端放置防滑槽。

Chamfering processing is necessary for stone external corner, and an anti-skidding slot should be placed in front of the steps.

微信扫码，了解更多关于"楼梯构造、画法"的知识

微信扫码，观看"石材踏步|混凝土楼梯|"三维节点动图

微信扫码，观看"石材踏步|钢结构楼梯|"三维节点动图

石材踏步　|钢结构楼梯|
Stone Step　| Steel Structure Staircase |

182P ↘　183P ↘

地坪工艺节点　DETAILS OF FLOOR PROCESSING TECHNIQUES

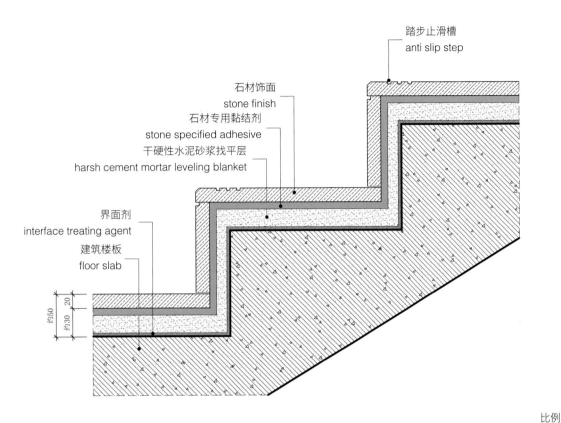

scale: 1:5

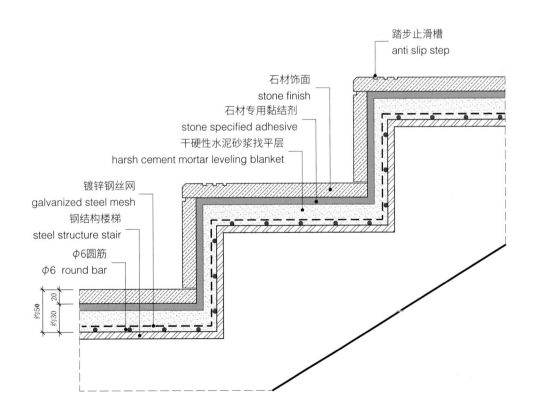

节点图　DETAIL

木地板踏步　|混凝土楼梯|
Wood Floor Step　| Concrete Staircase |

184P ↗　185P ↗

微信扫码，观看"木地板踏步 | 混凝土梯 |"三维节点动图

微信扫码，观看"木地板踏步 | 钢结构楼梯 |"三维节点动图

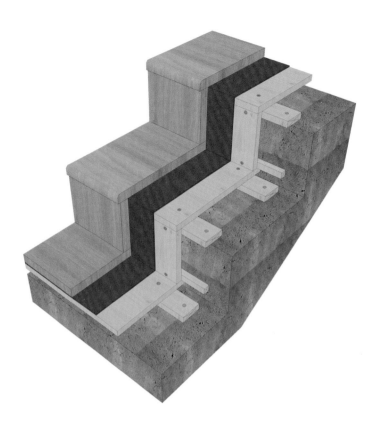

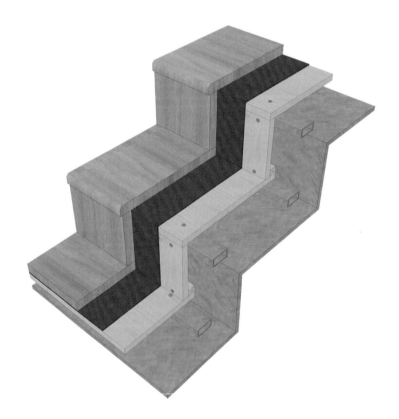

木地板踏步　|钢结构楼梯|
Wood Floor Step　| Steel Structure Staircase |

184P ↘　185P ↘

三维图　PERSPECTIVE

地坪工艺节点　DETAILS OF FLOOR PROCESSING TECHNIQUES

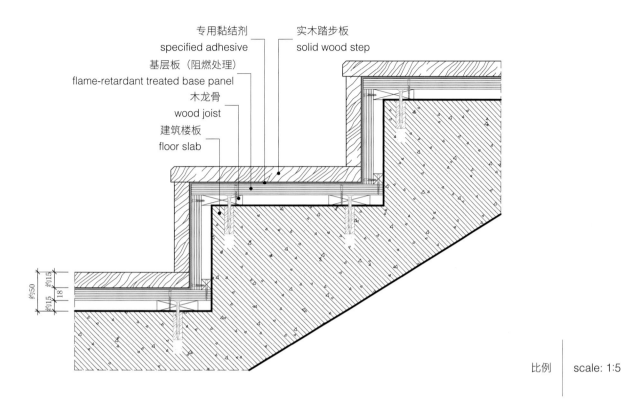

节点图　DETAIL

地毯踏步　|混凝土楼梯|
Carpet Step　| Concrete Staircase |

186P ↗　187P ↗

微信扫码，观看"地毯踏步
|混凝土楼梯|"三维节点
动图

微信扫码，观看"地毯踏步
|钢结构楼梯|"三维节点
动图

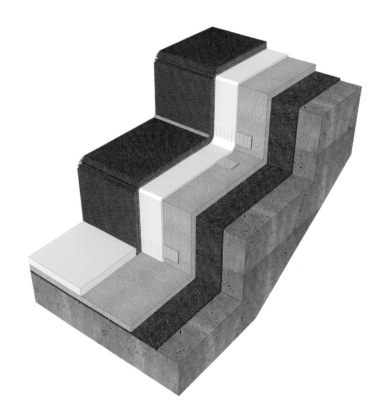

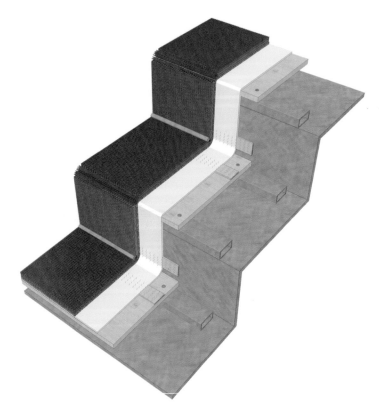

地毯踏步　|钢结构楼梯|
Carpet Step　| Steel Structure Staircase |

186P ↘　187P ↘

三维图　PERSPECTIVE

地坪工艺节点　DETAILS OF FLOOR PROCESSING TECHNIQUES

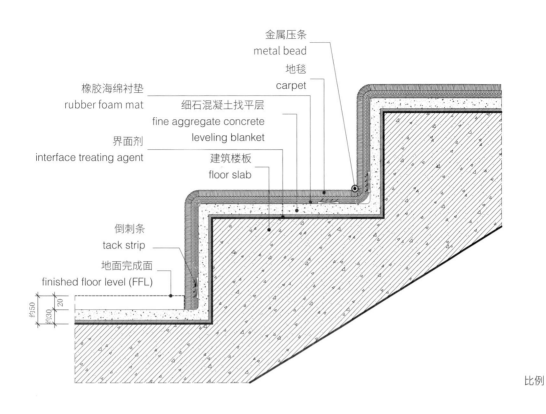

scale: 1:5

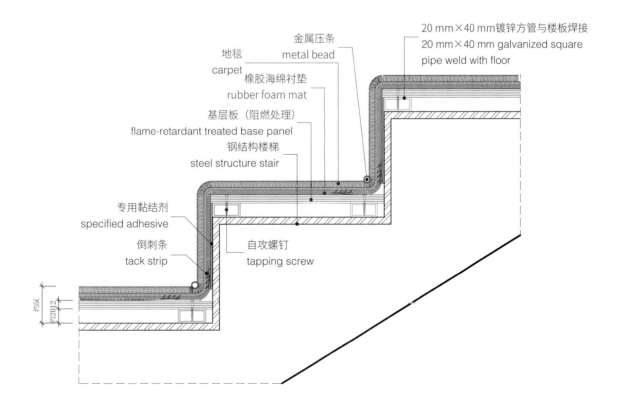

节点图　DETAIL

石材踏步　　|有灯带|
Stone Step　　| With Lamp Belt |

188P ↗　　189P ↗

重点 / KEY POINTS

注意灯带规格能否保证暗藏，不要暴露反光。

Pay attention to whether the specification of the lamp belt can be hidden and there should be no expose reflection.

微信扫码，观看"石材踏步 | 有灯带 |"三维节点动图

微信扫码，观看"木地板踏步 | 有灯带 |"三维节点动图

木地板踏步　　|有灯带|
Wooden Floor Step　　| With Lamp Belt |

188P ↘　　189P ↘

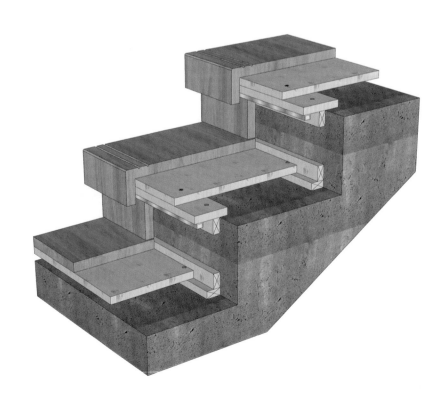

地坪工艺节点　DETAILS OF FLOOR PROCESSING TECHNIQUES

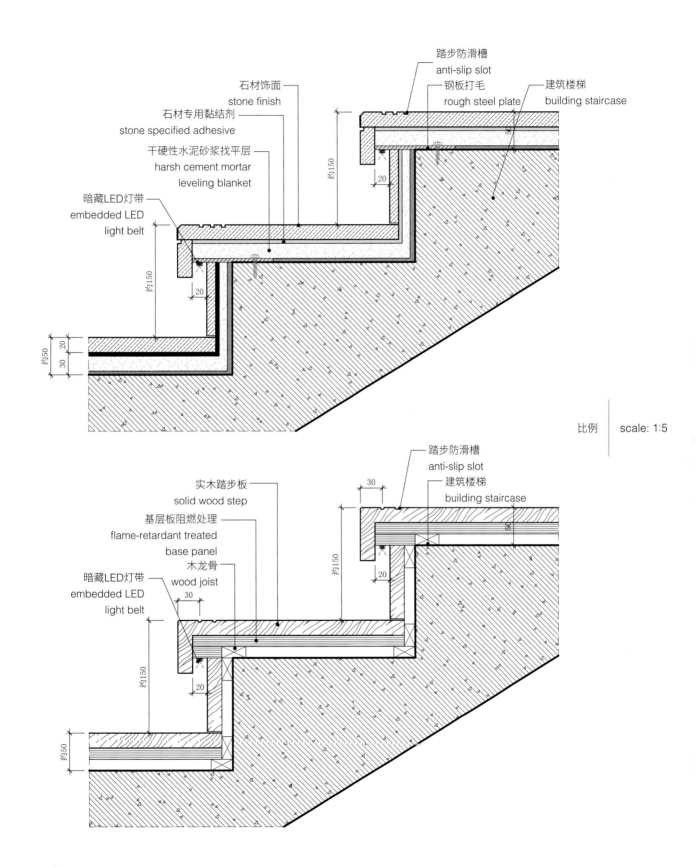

比例　scale: 1:5

节点图　DETAIL

4

门工艺节点
DETAILS OF DOOR PROCESSING TECHNIQUES

　　本篇主要选取了几种常见的门的工艺做法，对门的基本构造、所用五金进行了分析。需要提醒设计师的是：由于装饰市场的发展，门基本可以理解为是定加工的成品，其加工制作都是由专业工厂制作，现场进行安装。但是由于各个工厂的加工习惯、选材、成本等的不同，门、门框的制作工艺肯定不尽相同，所以设计师不必过度在意成品门的内部构造如何。

　　在装饰设计中，门是比较特殊又非常重要的一项。它是视觉的焦点，同时又有着频繁的使用需求，在特殊的部位还有着严格的规范要求。门的分类方式也有很多，比如按开门方式有单开门、双开门、折叠门等；按材质有玻璃、木门、金属门等；按规范功能有防火门、常规门等。正是由于门体系比较复杂，设计师更应该了解不同的空间、功能应该选用什么形式的门，以及配合什么样的五金。在满足基本使用功能的前提下，再去进行装饰形态等美学上的思考。

地弹簧玻璃门
Floor Spring Glass Door

192P / 193P

重点 / KEY POINTS

地弹门用地埋式门轴弹簧，门扇可以双向开启。

Spring door could be pushed or pulled by assembling underground pivot spring.

微信扫码，了解更多关于"门控五金"的知识

微信扫码，观看"地弹簧玻璃门"三维节点动图

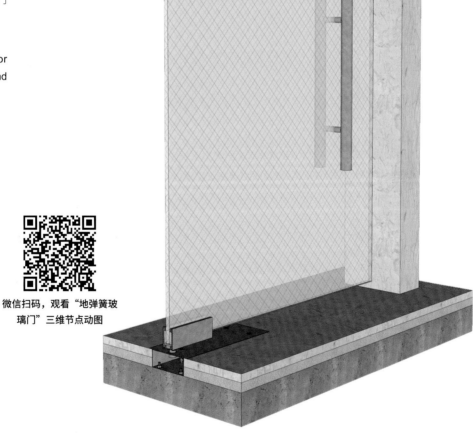

三维图　PERSPECTIVE

门工艺节点　DETAILS OF DOOR PROCESSING TECHNIQUES

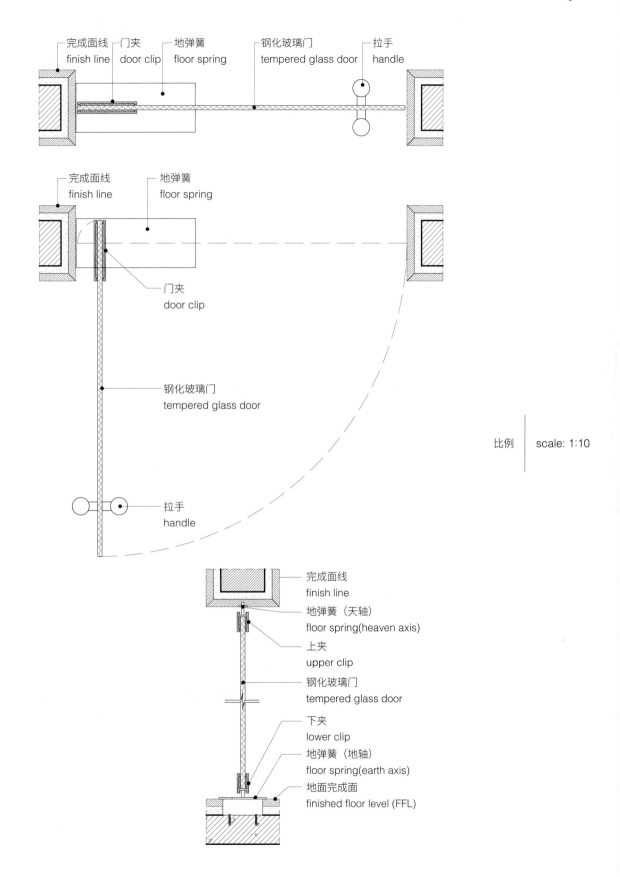

节点图　DETAIL

玻璃铰链门　|固定玻璃|
Glass Hinge Door　| Fixed Glass |

194P / 195P

微信扫码，观看"玻璃较链门|固定玻璃|"三维节点动图

门工艺节点　DETAILS OF DOOR PROCESSING TECHNIQUES

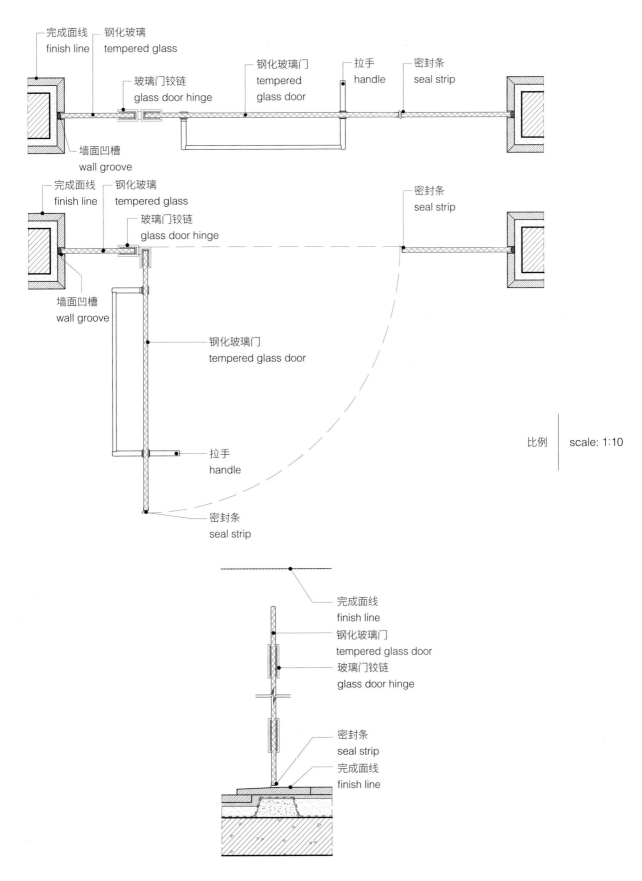

节点图　DETAIL

玻璃铰链门　|固定墙面|
Glass Hinge Door　| Fixed Wall |

196P / 197P

微信扫码，观看"玻璃较链门|固定墙面|"三维节点动图

三维图　PERSPECTIVE

门工艺节点　DETAILS OF DOOR PROCESSING TECHNIQUES

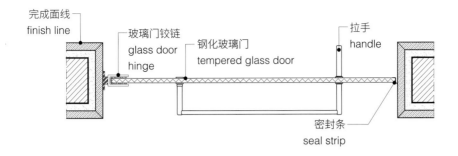
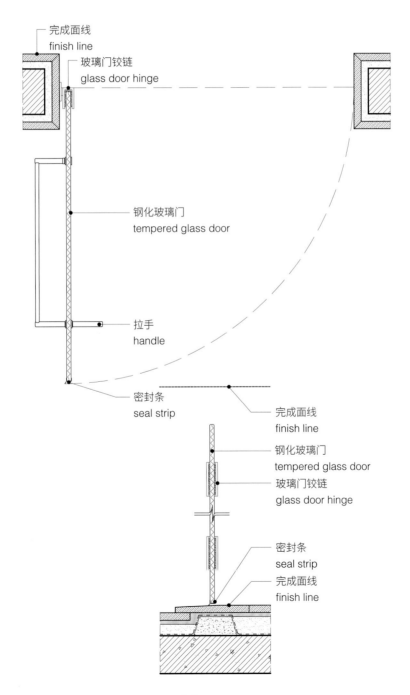

比例　scale: 1:10

节点图　DETAIL

双开门
Double Door

198P / 199P

微信扫码，了解更多关于
"门、门套构造"的知识

微信扫码，观看"双开门"
三维节点动图

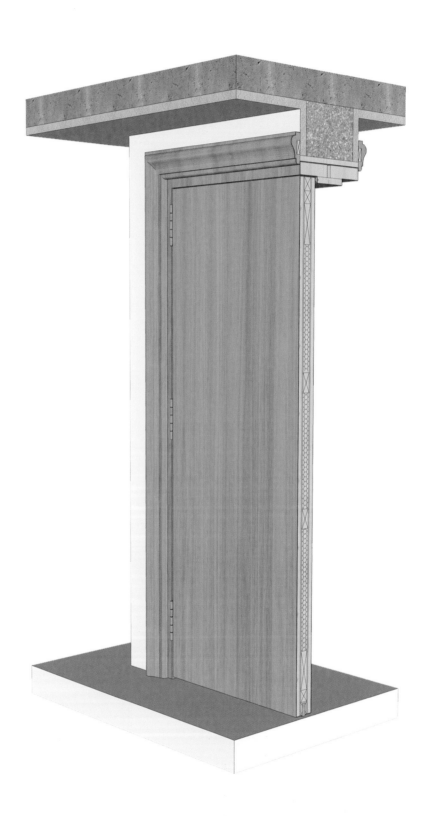

门工艺节点　DETAILS OF DOOR PROCESSING TECHNIQUES

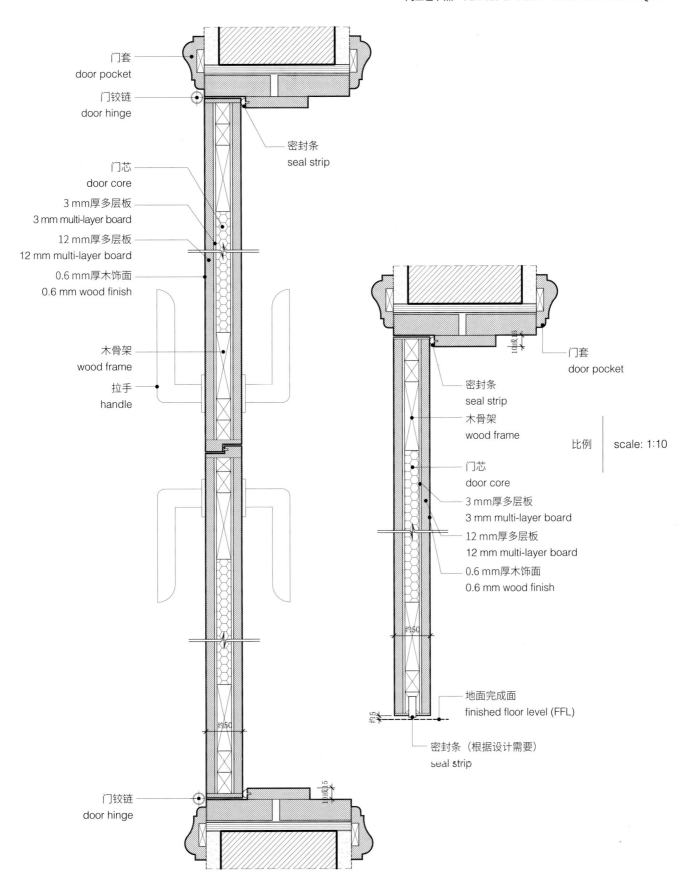

节点图　DETAIL

单开门
Single Door

200P / 201P

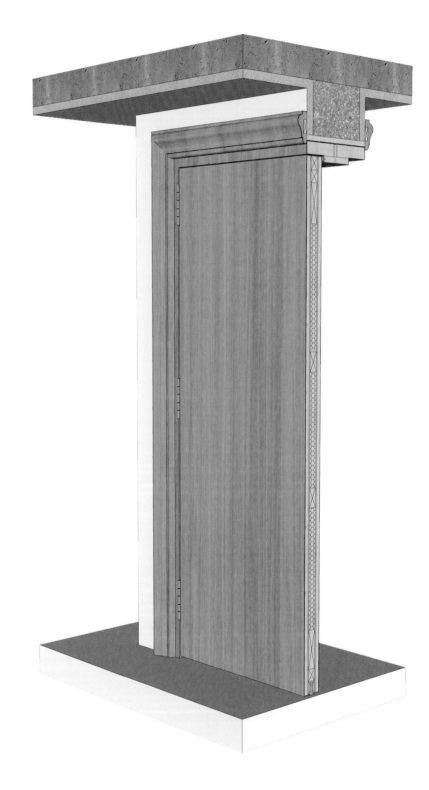

微信扫码，观看"单开门"
三维节点动图

三维图　PERSPECTIVE

门工艺节点 DETAILS OF DOOR PROCESSING TECHNIQUES

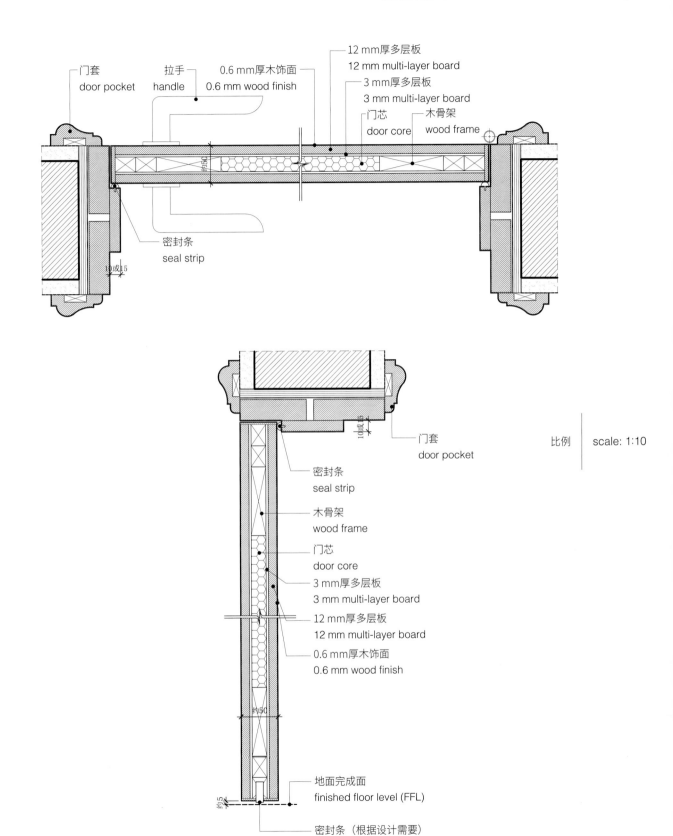

比例 scale: 1:10

节点图 DETAIL

暗藏移门
Hidden Door

202P / 203P

微信扫码，了解更多关于"暗藏移门"的知识

微信扫码，观看"暗藏移门"三维节点动图

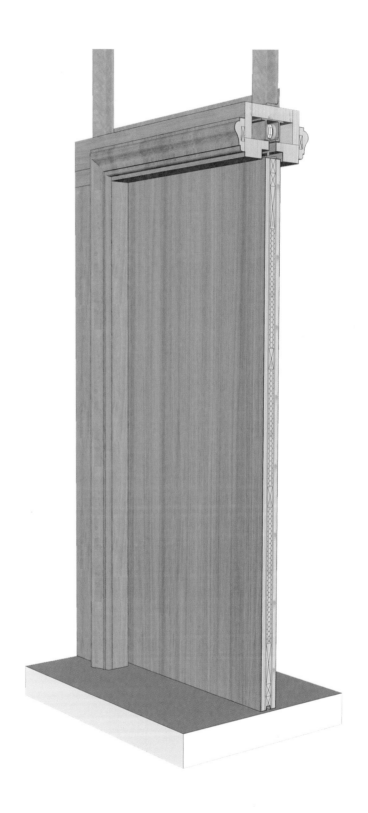

门工艺节点　DETAILS OF DOOR PROCESSING TECHNIQUES

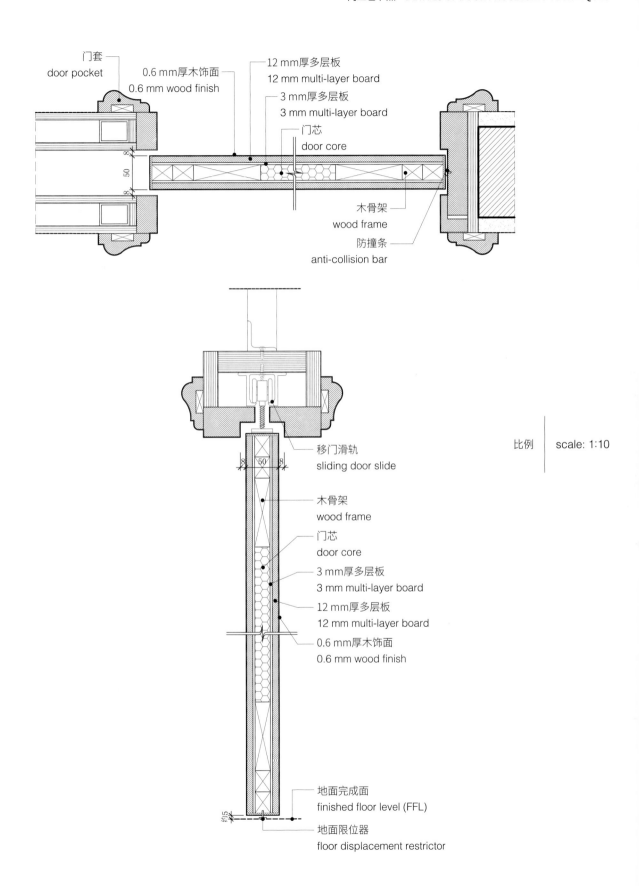

节点图　DETAIL

贴墙明装移门
Wall-mounted Sliding Door

204P / 205P

重点 / KEY POINTS

需要根据设计方案考虑墙体在立面上的收口关系。

It is necessary to consider the closure relationship of the wall on the elevation according to the design scheme.

微信扫码，了解更多关于"移门"的知识

微信扫码，观看"贴墙明装移门"三维节点动图

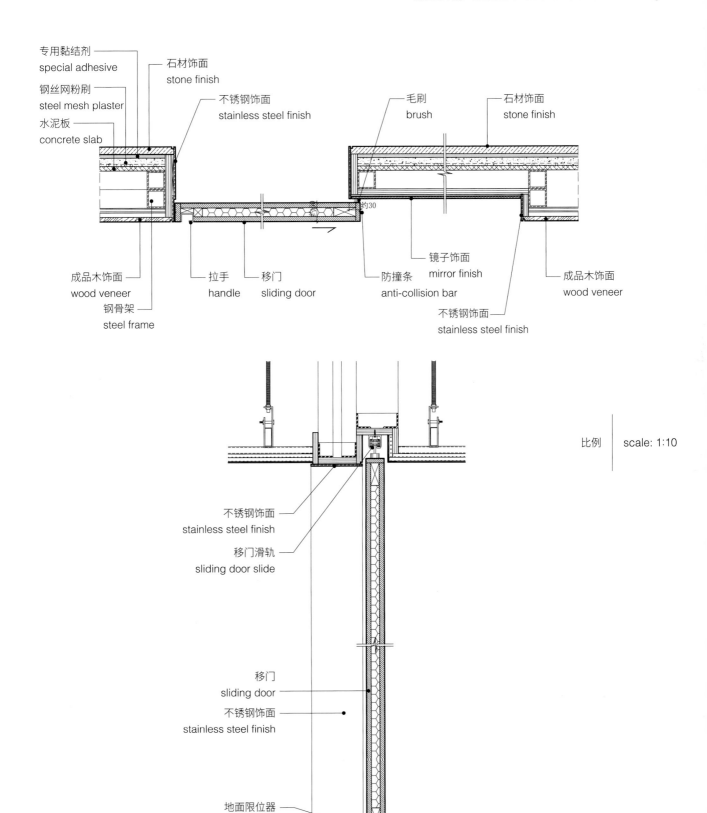

节点图　DETAIL

同向联动移门
Co-directional Linkage Sliding Door

206P / 207P

重点 / KEY POINTS

门扇尺寸注意预留门框重叠部分，选择适配联动移门的门五金。

Pay attention to reserve the overlapping part of the door frame and select the door hardware suitable for the co-directional linkage sliding door.

微信扫码，观看"同向联动移门"三维节点动图

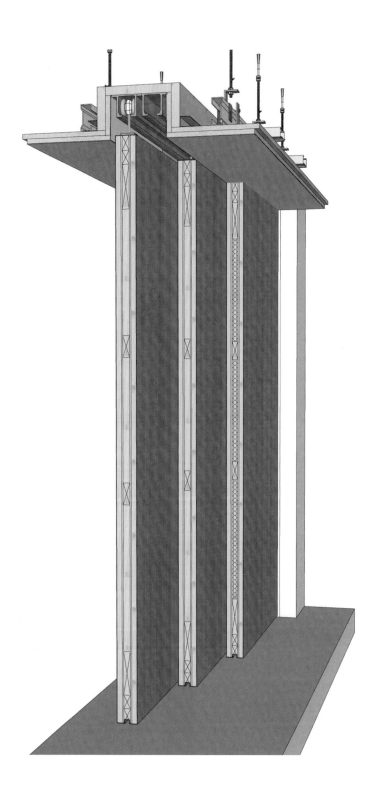

三维图 PERSPECTIVE

门工艺节点　DETAILS OF DOOR PROCESSING TECHNIQUES

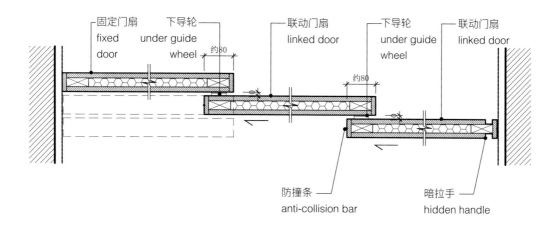

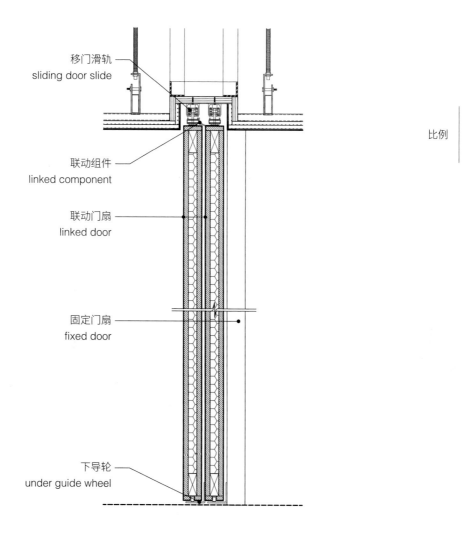

比例　scale: 1:10

节点图　DETAIL

电动玻璃移门
Electric Glass Sliding Door

208P / 209P

重点 / KEY POINTS

了解电机尺寸并判断安装空间是否足够，和供应商协调检修需求及检修方式。

Knowing the size of the motor, determining whether the installation space is enough, and coordinating with the supplier on maintenance requirements and maintenance methods are necessary.

微信扫码，观看"电动玻璃移门"三维节点动图

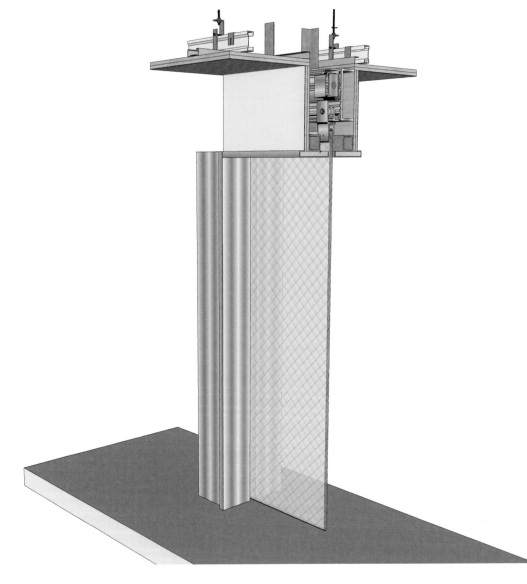

三维图　PERSPECTIVE

门工艺节点　DETAILS OF DOOR PROCESSING TECHNIQUES

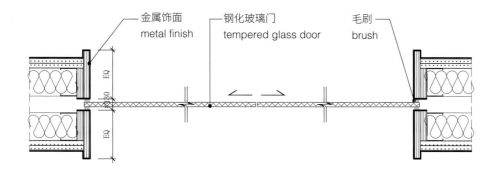

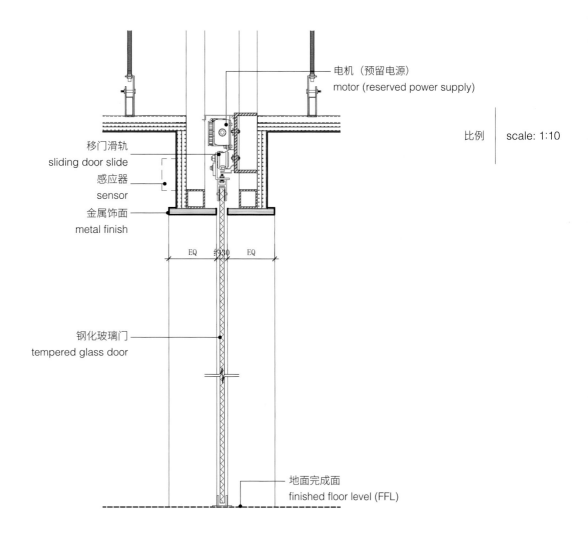

scale: 1:10

节点图　DETAIL

不锈钢电梯门套
Stainless Steel Elevator Door Cover

210P ↗   211P ↗

重点 / KEY POINTS

注意和原有电梯门套的收口关系；如果建筑预留电梯门洞尺寸较小，需要控制装饰完成面尺寸，相应的内部构造可以简化。

Pay attention to the closure relationship with the original elevator door cover; if the size of the reserved elevator door hole is small, the dimension of the decorative finish surface needs to be controlled, and the corresponding internal structure can be simplified.

微信扫码，观看"不锈钢电梯门套"三维节点动图

微信扫码，了解更多关于"电梯门套"的知识

微信扫码，观看"石材电梯门套"三维节点动图

石材电梯门套
Stone Elevator Door Cover

210P ↘   211P ↘

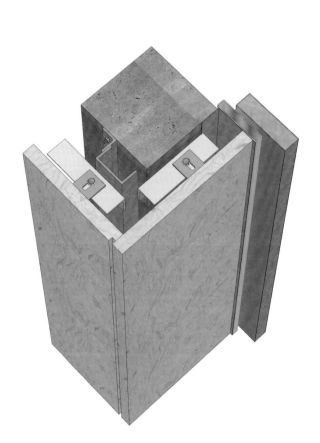

门工艺节点　DETAILS OF DOOR PROCESSING TECHNIQUES

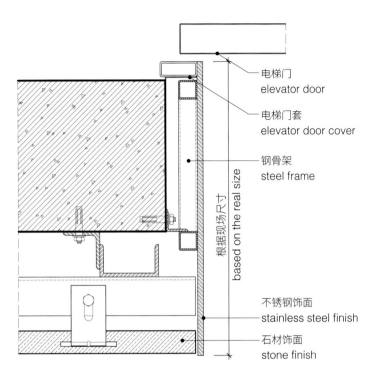

- 电梯门 elevator door
- 电梯门套 elevator door cover
- 钢骨架 steel frame
- 不锈钢饰面 stainless steel finish
- 石材饰面 stone finish

根据现场尺寸 based on the real size

比例　scale: 1:10

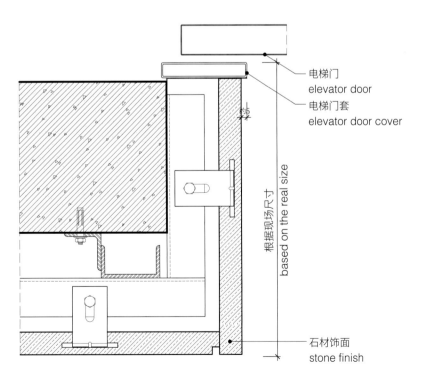

- 电梯门 elevator door
- 电梯门套 elevator door cover
- 石材饰面 stone finish

根据现场尺寸 based on the real size

节点图　DETAIL

钢制单开防火门
Steel Single Opened Fire-proof Door

212P / 213P

重点 / KEY POINTS

防火门由专业厂家设计生产，设计师可以在规范允许的范围内，在防火门上进行一些装饰处理，但需要和厂家进行沟通。

Fire-proof doors are designed and manufactured by professional manufacturers. Designers can make some decorative treatment on fire-proof doors within the allowable scope of codes, but the communication with manufacturers is needed.

微信扫码，了解更多关于"防火门"的知识

微信扫码，观看"钢制单开防火门"三维节点动图

三维图　PERSPECTIVE

门工艺节点 DETAILS OF DOOR PROCESSING TECHNIQUES

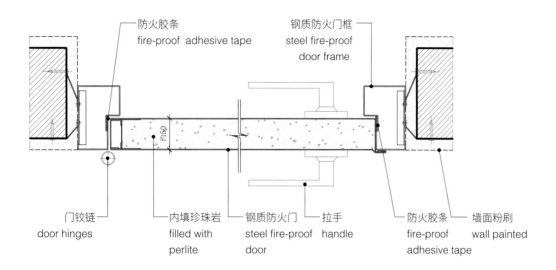

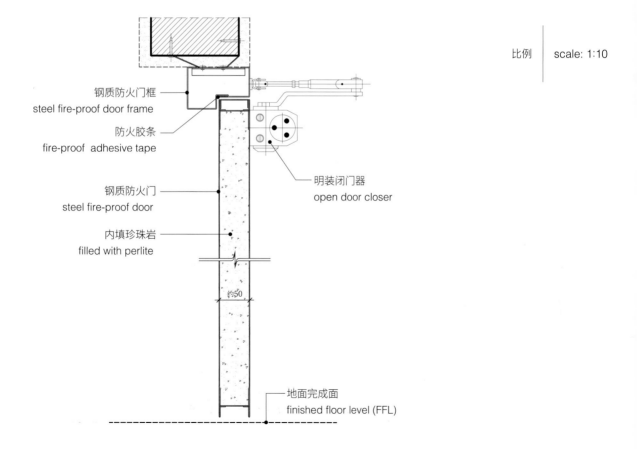

scale: 1:10

节点图 DETAIL

钢制子母防火门
Steel Dobule Fire -proof Door

214P / 215P

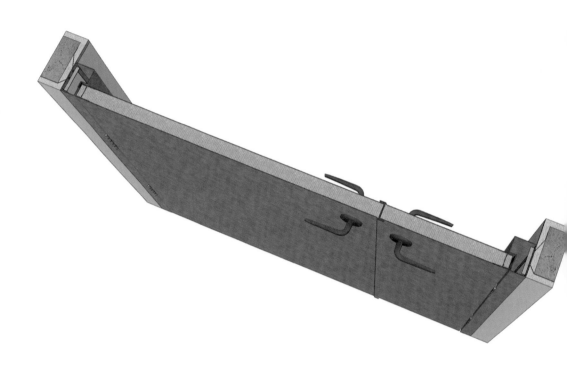

微信扫码，观看"钢制子母防火门"三维节点动图

门工艺节点　DETAILS OF DOOR PROCESSING TECHNIQUES

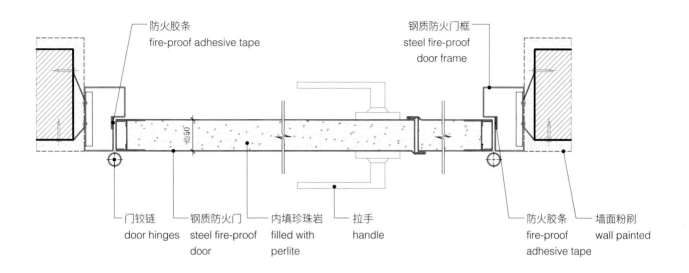

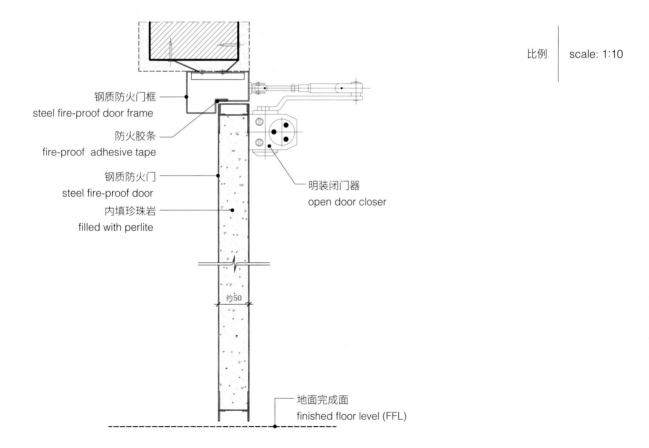

比例　scale: 1:10

节点图　DETAIL

木制单开防火门
Wooden Single Fire -proof Door

216P / 217P

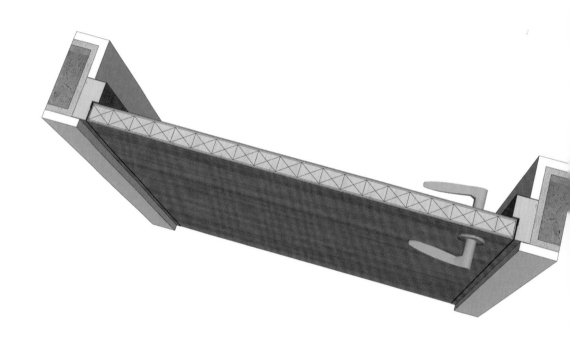

微信扫码，观看"木质单开
防火门"三维节点动图

三维图　PERSPECTIVE

门工艺节点　DETAILS OF DOOR PROCESSING TECHNIQUES

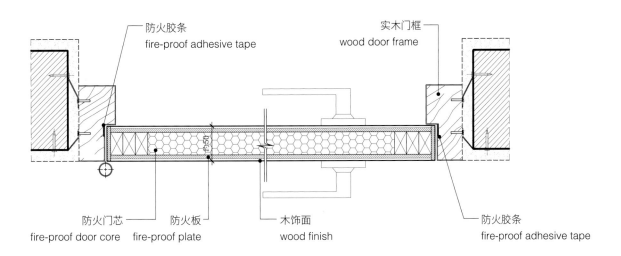

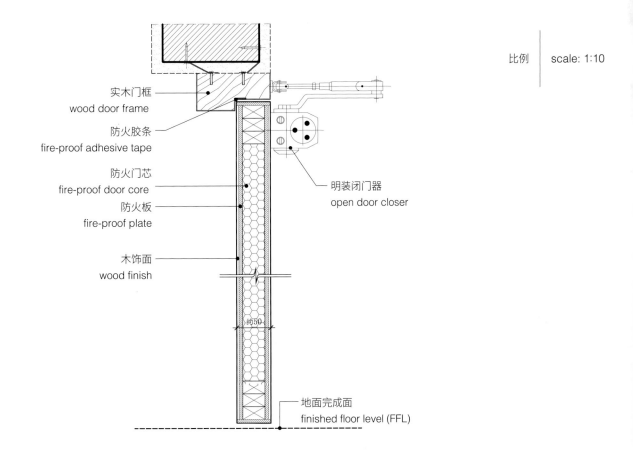

比例　scale: 1:10

节点图　DETAIL

木制子母防火门
Wooden Double Fire -proof Door

218P / 219P

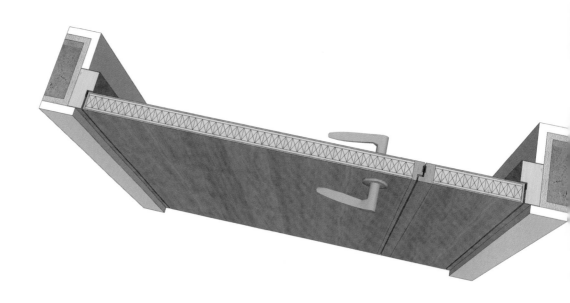

微信扫码，观看"木质子母
防火门"三维节点动图

门工艺节点　DETAILS OF DOOR PROCESSING TECHNIQUES

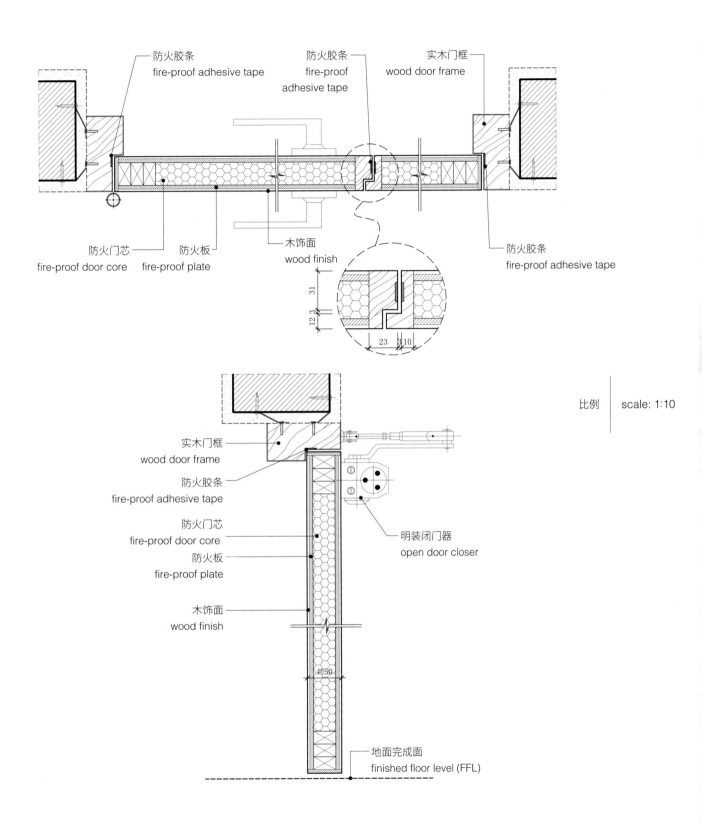

比例　scale: 1:10

节点图　DETAIL

常开防火门
Frequently Opened Fire -proof Door

220P / 221P

重点 / KEY POINTS

因为所处部位和功能的需要，这类防火门需要在平时维持开启状态，火灾发生时关闭。为了美观，可以在防火门开启状态时做出门扇和墙面嵌平的效果。

Due to the location and function, such fire-proof doors need to be kept open in ordinary time, but closed when fire occurs. For beauty, when the fire-proof door is open, the door and the wall can be made to a flat effect.

微信扫码，观看"常开防火门"三维节点动图

门工艺节点　DETAILS OF DOOR PROCESSING TECHNIQUES

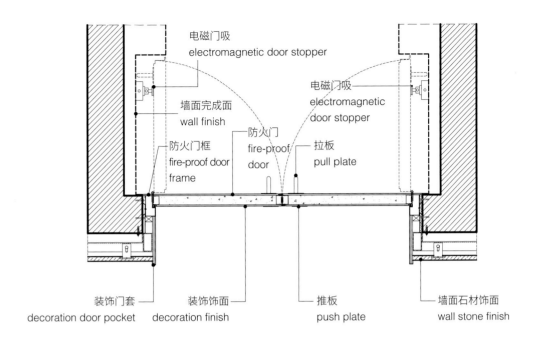

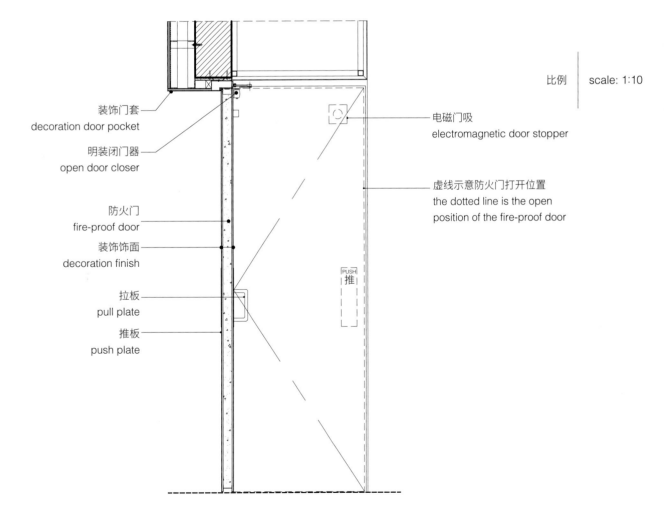

scale: 1:10

节点图　DETAIL

装饰暗门　|双道门|
Decorative Hidden Door　| Double door |

222P / 223P

重点 / KEY POINTS

防火门需要做成隐藏暗门时，可采用原有防火门不动，在防火门外部另加装饰暗门的处理手法。适用于机房管井等部位，需要有一定的墙面完成面厚度。

If there are hidden door requirements on fire-proof doors, the original fire doors do not need to be moved. The method by adding decorative hidden doors outside the fire-proof doors is enough. This method is suitable for the machine room, pipe well and other parts, and a certain thickness of wall finish is needed.

微信扫码，观看"装饰暗门|双道门|"三维节点动图

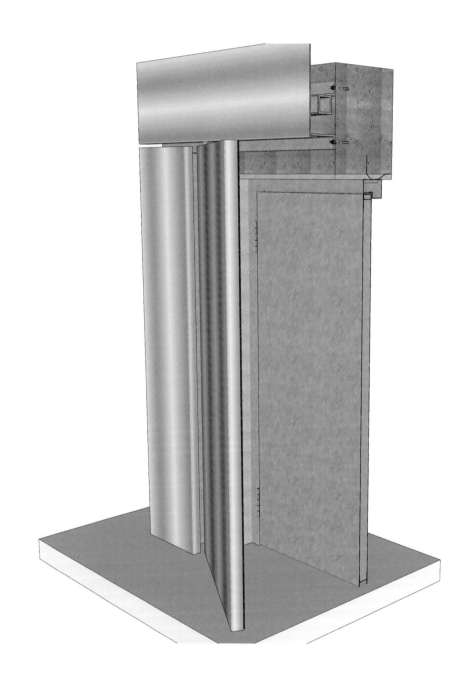

三维图　PERSPECTIVE

门工艺节点　DETAILS OF DOOR PROCESSING TECHNIQUES

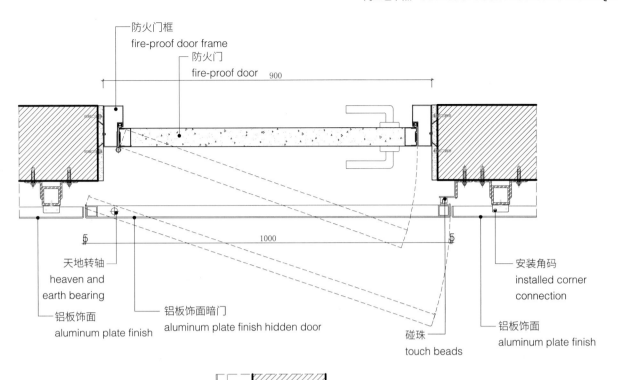

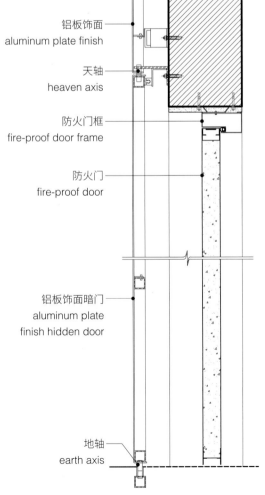

节点图　DETAIL

scale: 1:10

石材暗门
Stone Hidden Door

224P / 225P

重点 / KEY POINTS

石材暗门在装饰设计中多应用于消火栓暗门，消火栓暗门需要做到开门见消火栓箱，不能有遮挡，同时门的开启角度理论上应该达到160°。

In decorative design, stone hidden door is often used in fire hydrant where the fire hydrant boxes must be seen without barrier immediately the door is open. And the opening angle of the door should reach 160° theoretically.

微信扫码，了解更多关于"消火栓暗门"的知识

微信扫码，观看"石材暗门"三维节点动图

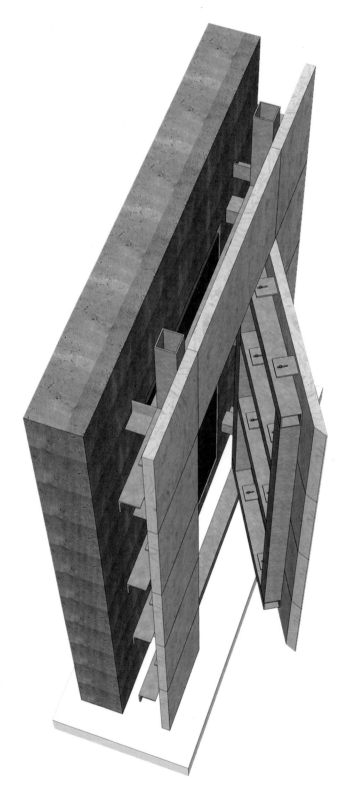

门工艺节点 DETAILS OF DOOR PROCESSING TECHNIQUES

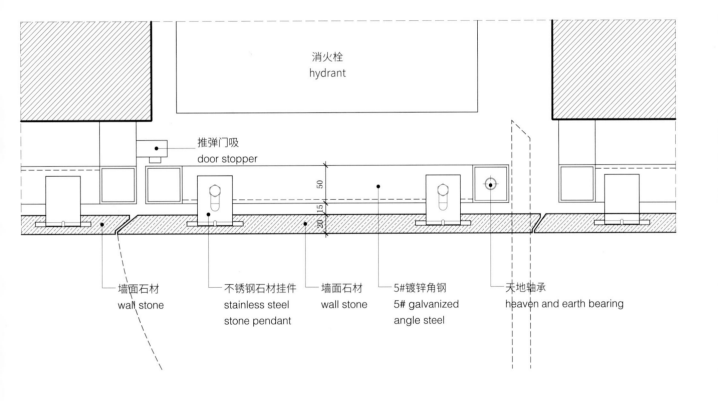

节点图　DETAIL

scale: 1:10

门工艺节点　DETAILS OF DOOR PROCESSING TECHNIQUES

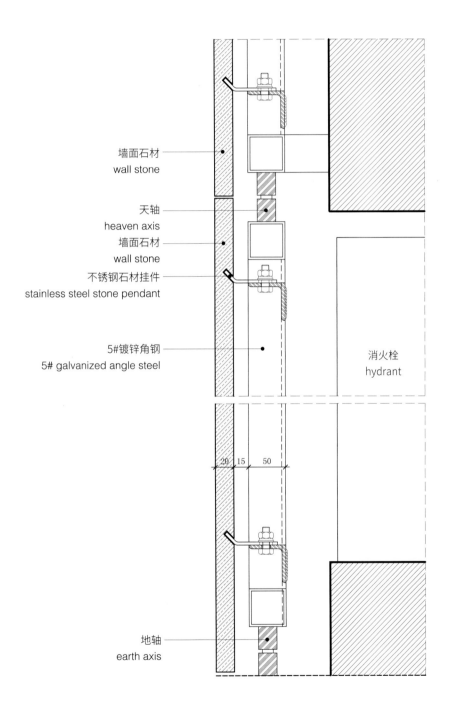

节点图　DETAIL